Web 3.0

The book underscores AI's transformative impact on reshaping physical, digital, and biological boundaries, converging with technologies like robotics, IoT, 3D printing, genetic engineering, and quantum computing—termed Web 3.0 Industrial Revolution. This global revolution integrates advanced production techniques beyond connected machines, extending into gene sequencing, nanotechnology, renewable energies, and quantum computing. The book's main goals include providing a collaborative platform for academia and industry researchers to share contributions and shape the future through knowledge exchange. Recognizing recent progress driven by increased computing power, it highlights the positive impact of digital technology—AI, IoT, AR/VR, Additive Manufacturing, CPS, cloud computing, and robotics—on industrial efficiency and quality.

- Revolutionary AI Fusion: AI revolutionizes by blending physical, digital, and biological boundaries through cutting-edge technologies like robotics, IoT, 3D printing, genetic engineering, and quantum computing.
- Global Manufacturing Cooperation: AI creates a collaborative landscape where virtual and physical systems flexibly cooperate on a global scale.
- AI's Diverse Impact: Beyond smart machines, AI drives breakthroughs in gene sequencing, nanotechnology, renewable energies, and quantum computing, distinguishing it from prior industrial revolutions.
- Progress and Digital Interface: Recent progress, powered by computing advancements, boosts industrial efficiency. The digital technology interface (AI, IoT, AR/VR, 3D Printing, CPS, CC, Robotics) significantly impacts industrial performance.

In conclusion, AI spearheads a transformative revolution, redefining the boundaries of the physical, digital, and biological realms. The fusion of AI with Web 3.0 Industrial Revolution, integrating advanced production techniques and global manufacturing cooperation, has surpassed past industrial shifts. The book aims to be a collaborative platform for academia and industry researchers, fostering knowledge exchange to shape the future. In AI-driven manufacturing within Web 3.0, a paradigm shift envisions maximum output with minimal resource use. Coupled with 'Digital Reality,' it transforms business practices, consumer behavior, and employment dynamics, redistributing wealth toward innovation and technology.

Web 3.0
The Next Generation's Internet and Understanding the Concept of the Metaverse

Edited by
Prabhat Kumar Srivastav
Prateek Singhal
Basudeo Singh Roohani
Nitin Sharma

CRC Press
Taylor & Francis Group
Boca Raton London New York

CRC Press is an imprint of the
Taylor & Francis Group, an **informa** business

Designed cover image: Shutterstock

First edition published 2024
by CRC Press
2385 NW Executive Center Drive, Suite 320, Boca Raton FL 33431

and by CRC Press
4 Park Square, Milton Park, Abingdon, Oxon, OX14 4RN

CRC Press is an imprint of Taylor & Francis Group, LLC

ISBN: 978-1-032-60987-4 (hbk)
ISBN: 978-1-032-60988-1 (pbk)
ISBN: 978-1-003-46141-8 (ebk)

DOI: 10.1201/9781003461418

Typeset in Sabon
by MPS Limited, Dehradun

Contents

Contributors

Himanshi Agrawal
Meerut Institute of Engineering &
  Technology
Meerut, Uttar Pradesh, India

Taslima Ahmed
Department of Electronics
  Communication Engineering
IIMT College of Engineering
Greater Noida, Uttar Pradesh, India

Kunwar Babar Ali
Department of Computer Science &
  Engineering (AI)
Meerut Institute of Engineering &
  Technology
Meerut, Uttar Pradesh, India

Dr. Neeti Bansal
Meerut Institute of Engineering &
  Technology
Meerut, Uttar Pradesh, India

Shivani Chaudhary
Student of Bachelor of
  Computer Applications
IMS-Ghaziabad University
  Courses Campus
Ghaziabad, Uttar Pradesh, India

Ashish Dixit
Department of CSE
Ajay Kumar Garg Engineering
  College
Ghaziabad, Uttar Pradesh, India

Lav Kumar Dixit
RDEC
Ghaziabad, Uttar Pradesh, India

Khushi Garg
Student of Bachelor of Computer
  Applications
IMS-Ghaziabad University Courses
  Campus
Ghaziabad, Uttar Pradesh, India

Dr. Ajay Kumar Gupta
IIMT College of Engineering
Greater Noida, Uttar Pradesh, India

Purnima Gupta
IMS-Ghaziabad University Courses
  Campus
Ghaziabad, Uttar Pradesh, India

Dr. Rolly Gupta
Department of Computer Science
  Engineering
SRMIST
Modi Nagar, Ghaziabad,
  Uttar Pradesh, India

Zaiba Ishrat
Department of Electronics
  Communication Engineering
Meerut Institute of Technology
Meerut, Uttar Pradesh, India

Jaishree Jain
Department of CSE
Ajay Kumar Garg Engineering
  College
Ghaziabad, Uttar Pradesh, India

Dr. K. Rama Krishna
GNIOT Engineering Institute
Greater Noida, Uttar Pradesh, India

Anuj Kumar
Department of CSE
Ajay Kumar Garg Engineering
  College
Ghaziabad, Uttar Pradesh, India

Bhupendra Kumar
IIMT University
Meerut, Uttar Pradesh, India

Dr. Jaideep Kumar
Department of CSE (IOT)
Raj Kumar Goel Institute of
  Technology
Ghaziabad, Uttar Pradesh, India

Manish Kumar
Department of CSE
Ajay Kumar Garg Engineering
  College
Ghaziabad, Uttar Pradesh, India

Amrita Kumari
Quantum University
Roorkee, Uttrakhand, India

Dr. Seema Malik
Department of ECE
Raj Kumar Goel Institute of
  Technology
Ghaziabad, Uttar Pradesh, India

Ashish Pandey
IMSEC
Ghaziabad, Uttar Pradesh, India

Dr. Shailendra Prakash
IIMT College of Engineering
Greater Noida, Uttar Pradesh, India

Dr. Lalit Kumar Sagar
Department of Computer Science
  Engineering
SRMIST
Modi Nagar, Ghaziabad,
  Uttar Pradesh, India

Ashish Saini
Quantum University
Roorkee, Uttrakhand, India

Kewal Krishan Sharma
IIMT University
Meerut, Uttar Pradesh, India

Vikas Sharma
IIMT University
Meerut, Uttar Pradesh, India

Prof Swati S Sherekar
P.G. Department of Computer
  Science
SGBAU
Amravati, Maharashtra, India

Aswani Kumar Singh
Soft-tech Development Solution,
  Pt. Deen Dayal Upadhyay
Chandauli, Uttar Pradesh, India

Ramander Singh
RD Engineering College
Ghaziabad, Uttar Pradesh, India

Dr. Priyanka C Tikekar
Bharatiya Mahavidyalaya Amravati
Maharashtra, India

Abhishek Tyagi
Department of Computer Science
  and Engineering
IIMT College of Engineering
Greater Noida, Uttar Pradesh, India

Shekhar Tyagi
Department of Computer Science
  and Engineering
KIET Group of Institutions
Ghaziabad, Uttar Pradesh, India

Tarun Kumar Vashishth
IIMT University
Meerut, Uttar Pradesh, India

About the editors

Dr. Prabhat Kumar Srivastava is currently working at the IMS Engineering College as a professor in the department of CSE. He has 23 years of teaching and research experience. He has completed his PhD from SHUATS Allahabad and MTech from UPTU Lucknow. His area of research is **soft computing, machine learning, and fuzzy theory.** He has given valuable inputs as a guide to 25+ M.Tech students and 150+ B.Tech students in his career so far. He has published 15 Indian patents and 2 international patents with grants. He has published 17 conference papers and 51 journal papers in which 3 are in SCI journal and some are in reputed journals and IEEE conferences. He has worked as a reviewer/editorial board member for several reputed international journals. He has also served as a speaker, session chair, or co-chair at various national and international conferences. He earned numerous international certifications, such as data analytics, machine learning, and blockchain. He is currently associated with the Artificial Intelligence Foundation as a technical advisor of the Student Chapter. He is an active member of the ACM, IEEE, and CSI societies.

Mr. Prateek Singhal is an assistant professor in the Department of Computer Science at CHRIST (Deemed to be University) NCR-Ghaziabad, U.P., India. He is pursuing a PhD degree in medical imaging from the Maharishi University of Information Technology, Lucknow, India. He has more than four years of experience in research and teaching. He has published several research articles in SCI/SCIE/Scopus journals and conferences of high repute. He has also authored a book on cloud computing. He has various national and international patents and some are granted. He holds contributions in IEEE, Elsevier, etc., reputed journals. He is on the team of the research advisory members at his present institute.

His current areas of interest include **image processing, medical imaging, human computation interface, neuro-computing, and Internet of Things.** He is an active member of ACM and IEEE.

Mr. Basudeo Singh Roohani is currently working in the department of Computer Science and Engineering at the IMS Engineering College in Ghaziabad. He completed a B.Tech in computer science and engineering from the Bundelkhand Institute of Engineering & Technology (BIET), Jhansi; MTech in computer science from Uttar Pradesh Technical University (UPTU), Lucknow; and PhD(p) in computer science and engineering from Dr. A.P.J. Abdul Kalam Technical University, Lucknow. He has research interest in the fields of **design and analysis of algorithms, machine learning, and deep learning.** He has 12 years of administrative experience as head of the Computer Science and Engineering Department out of 22 years of experience at various reputed institutions, to monitor all operations and activities within the department. He has attended 13 FDPs, seminars, workshops, conferences, and training programs. He also presented nine research papers at national and international conferences. He has published seven patents. He has published three book chapters. He has also published seven research papers in reputed journals. He has also worked as a conference paper reviewer.

Dr. Nitin Sharma, an assistant professor in the CSE department at IMS Engineering College, has 13 years of experience in teaching and research. His research interests include machine learning, deep learning, and data analytics. He has guided over 6 MTech and 150 BTech students' projects and holds four Indian patents. Dr. Sharmahas contributed more than 12 research papers in several international and national journals and conference proceedings of high repute. He has also attended various FDP/STTP and workshops on emerging technologies like Python, Power BI, and artificial intelligence. He has also completed 15 days of FDP on entrepreneurship from EDII, Ahmedabad. One international book chapter in *Big Data Analytics* is also in his name. He has a BTech in information technology from U.P.T.U, Lucknow, and a master's in computer science and engineering from Uttarakhand Technical University, Dehradun. Dr. Sharma has completed his doctoral research in computer science and engineering from Sunrise University, Alwar (Rajasthan).

Chapter 1

A study on artificial intelligence

An analysis of recent research and future predictions

Amrita Kumari and Ashish Saini

Department of Computer Science & Engineering, Quantum University, Roorkee, Uttrakhand, India

1.1 INTRODUCTION

Artificial intelligence (AI) is a field of computer science concerned with creating computer systems capable of doing activities that normally require human intelligence, such as decision-making, visual perception, speech recognition, and language translation. AI is based on the idea of creating machines that can "think" like humans and can learn from experience and progress over time. The term "artificial intelligence" marks the first significant step toward exploring how human-intelligent activities can be simulated by machines.

The global interest in AI was immediately arouse in initial 2016, when the world chess champion was defeated by AlphaGo, leading to a surge in AI-related research (Beam et al., 2023). The development of AI has not only generated significant economic benefits for humankind but has also positively impacted all aspects of life and propelled social development into a new era (Huynh-The et al., 2023). As a result, numerous scholars have focused on AI-related research since the late 20th century.

The term "AI" encompasses the science of simulating human intelligent behaviors using computers, including learning, decision-making, and judgment (Xu et al., 2021). AI is primarily a knowledge endeavor that obtains, analyzes, and investigates knowledge expression methods with the goal of imitating human intellectual operations. AI is a blend of computer science, psychology, philosophy, biology, logic, and other fields that has yielded outstanding results in a variety of applications.

AI is typically divided into two main classes: narrow or weak AI, designed to perform a specific task or set of tasks, and general or strong AI, which has been designed to accomplish any intellectual task that a human can. Currently, most AI applications fall under the narrow AI category, but researchers continue to work on developing more advanced AI systems.

Some common examples of AI applications include speech recognition software, image and video recognition, virtual personal assistants, natural language processing, and autonomous vehicles. AI is also used in industries such as finance, healthcare, manufacturing, and transportation to help streamline processes and improve efficiency.

AI is a rapidly developing field with a varied range of potential applications, but there are also concerns about the ethical and social implications of AI development, such as job displacement and algorithmic bias. As such, researchers and policymakers are working to develop guidelines and regulations to ensure that AI is developed and used in a responsible and ethical manner. AI has become a crucial element in social development and has revolutionized labor efficiency, reduced labor costs, optimized human resource structures, and created new job opportunities.

1.2 ARTIFICIAL INTELLIGENCE'S ORIGINATION AND EVOLUTION

Artificial intelligence has been around since ancient times, with stories and myths portraying intelligent machines and robots. The current development of AI, on the other hand, began in the 1950s, with the establishment of computer science as a subject of study.

The Dartmouth Conference, organized in 1956 by John McCarthy, Marvin Minsky, and other scholars, is regarded as the genesis of artificial intelligence as an area of study. The conference gathered together experts from several fields to discuss the possibility of developing machines that could simulate human intellect.

The early years of AI were marked by high expectations and significant advancements, including the development of the first AI program, the Logic Theorist, by Allen Newell and Herbert Simon in 1955, and the introduction of the term "artificial intelligence" by John McCarthy in 1956. However, progress was slower than anticipated, and by the 1970s, interest in AI had waned. In the 1980s and 1990s, AI experienced a resurgence, with the development of new techniques, such as expert systems, neural networks, and machine learning. These advances led to the creation of intelligent systems that could perform tasks such as speech recognition, computer vision, and natural language processing (NLP) (Baclic et al., 2020; Olveres, 2021).

Since 1993, AI has made several breakthroughs, with the neural network growing rapidly, largely thanks to the widespread adoption of the back propagation (BP) algorithm. The broad utilization of expert systems in large-scale environments has significantly cut industry costs and enhanced efficiency (Zhang & Chu, 2020). The PROSPECTOR expert system is one illustration of this that has been used to analyze mineral discoveries with a value of hundreds of millions of dollars (Reddy & Fields, 2022). Following this, researchers tried to create generic artificial intelligence programs, but they came across significant challenges and became stuck. Subsequently, the development of AI stalled once again. However, the 1997 success of "Deep Blue" propelled AI back into the spotlight, and advancements in computing power broke down previous bottlenecks, facilitating continued progress in enhanced learning that

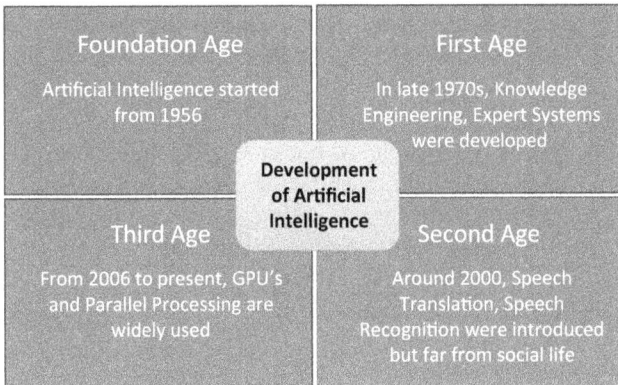

Foundation Age

Artificial Intelligence started from 1956

First Age

In late 1970s, Knowledge Engineering, Expert Systems were developed

Development of Artificial Intelligence

Third Age

From 2006 to present, GPU's and Parallel Processing are widely used

Second Age

Around 2000, Speech Translation, Speech Recognition were introduced but far from social life

Figure 1.1 Development of artificial intelligence.

is based on big data and deep learning (Chen et al., 2022). Custom processors, as well as the continued progress of graphic processing units (GPUs), have greatly expanded processing capability, laying the framework for AI's stratospheric rise. Figure 1.1 shows the development of AI.

The advancement of artificial intelligence has been a lengthy process, spanning over 70 years and involving various stages. The artificial neuron model was introduced in 1943, initiating the research era of artificial neural networks (Chhajer et al., 2022). During this period, international academic communities witnessed an upsurge in AI research and frequent academic exchanges. In the 1960s, connectionism and submissive techniques became less prevalent, leading to a slowdown in smart technology development. The 1970s saw the inception of backpropagation algorithm research, which was hindered by the high cost and computing power of computers, making it challenging to research and apply expert systems. Despite the obstacles, AI gradually made progress. By the 1980s, backpropagation neural networks gained wide recognition, and artificial neural network–based algorithm research developed rapidly. Computer hardware functionalities also enhanced, and the growth of the internet wedged the evolution of AI (Zhang & Lu, 2021). The development of mobile internet during the first decade of the twenty-first century gave rise to several artificial intelligence–based application scenarios. The year 2012 witnessed the proposal of deep learning, resulting in AI's breakthrough development, with the algorithm achieving significant advancements in speech and visual recognition (Olveres, 2021).

The exploration of how to enable computers to perform intelligent tasks exclusive to humans is the essence of artificial intelligence. Over the years, AI has made tremendous strides and had a significant impact on people's lifestyles (Zhang, 2022). The 21st century has seen the rapid growth of AI, with breakthroughs in areas such as deep learning, reinforcement learning, and robotics. AI is now being used in a wide range of applications, from virtual personal assistants to self-driving cars to medical diagnosis and treatment.

With AI being considered a crucial developmental strategy for countries worldwide, it now plays a crucial role in enhancing security and national competitiveness (Zhang et al., 2022). Several nations have developed beneficial policies to gain an advantage in the new era of global competition (Zhang & Lu, 2021). As a result, AI has emerged as a well-known research topic in science and technology, with major corporations such as Microsoft, IBM, and Google investing resources to it and employing it in an increasing variety of sectors. Human-computer interaction, cognition, machine learning, emotion detection, data storage, and decision-making abilities are all drawn by artificial intelligence.

As AI has evolved, so have concerns about its potential impact on society, such as job displacement and algorithmic bias (Owsley & Greenwood, 2022). Researchers and governments are attempting to address these concerns and ensure that artificial intelligence is developed and used responsibly and ethically.

1.3 FACILITATING FACTORS AND TECHNOLOGIES THAT DRIVE ARTIFICIAL INTELLIGENCE

There are some facilitating factors that drive AI. Some of them are mentioned below and also depicted in Figure 1.2.

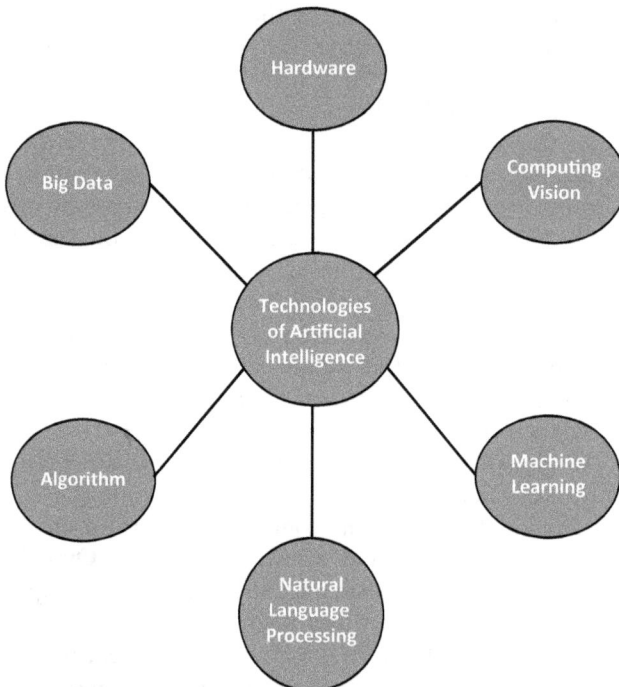

Figure 1.2 Facilitating factors that drive artificial intelligence.

1.3.1 Big data

AI necessitates big data, which is a critical component for improving identification accuracy and rate. The Internet of Things' (IoT) introduction and widespread adoption has resulted in an exponential increase in the amount of data created, with a dramatically enhanced yearly growth rate. Aside from the increased amount of data points, the data dimensions have also increased. The accessibility of high-dimensional data makes AI advancement more inclusive and sufficient.

1.3.2 Machine learning

Machine learning is based on the use of an algorithm that improves performance through data learning. Prediction, classification, clustering, and dimensionality reduction are all challenges that machine learning can solve (Janiesch et al., 2021). Based on the learning process, machine learning can be divided into four categories: supervised learning, unsupervised learning, semi-supervised learning, and reinforcement learning (Janiesch et al., 2021).

Supervised learning involves training with labeled data for predicting the value of new data or its type. It can be further categorized into classification and regression, depending on the prediction results. Support vector machine (SVM) and linear discrimination are typical methods of supervised learning (Saranya et al., 2020). Regression deals with predicting the output of continuous values. For instance, it is possible to analyze and fit home price data using input sample data before using the resulting continuous curve to forecast house prices.

On the other hand, in order to determine if an image of a dog or a cat is based on a collection of qualities, classification focuses on predicting the consequences of discrete values. The outcome is a value of 1 or 0, depending on the case.

The major use of unsupervised learning is clustering, which involves classifying data based on distinct features without using tags. Unsupervised learning approaches that are commonly used include principal component analysis and K-clustering (Saranya et al., 2020). Principal component analysis is a statistical method that transforms correlated variables into uncorrelated variables via orthogonal transformation. Principal components are the modified variables. The primary aim is to substitute the initial connected indications with a collection of independent comprehensive indicators. K-clustering, on the other hand, is based on calculating the difference between data points using Euclidean distance (Ghazal, 2021). If a measurement is not possible, the distance must be transformed to a useful Euclidean distance.

On the other hand, semi-supervised learning combines supervised and unsupervised learning, combining labeled and unlabeled data in the learning process. Typically, there is significantly a larger amount of unlabeled data than labeled data. Semi-supervised learning is a promising concept, although it is not frequently applied in real-world settings. Self-training, semi-supervised support vector machines (S3VM), and graph-based semi-supervised learning are a few of the semi-supervised learning techniques that are frequently utilized.

The reinforcement learning method entails gaining rewards through environmental interaction and evaluating the effectiveness of actions based on reward levels. After that, the model is trained appropriately. Exploration and development must be balanced in reinforcement learning because humans must select the course of action that may result in the greatest reward while also learning new activities. Behaviorist Thorndike (Kosinski & Zaczek-Chrzanowska, 2007) proposed a useful rule in 1911, which laid the groundwork for reinforcement learning: individuals and animals tend to reinforce actions in comfortable environments and reduce movements in uncomfortable environments. Through trial-and-error training, the model is educated to identify the ideal operation and behavior to provide the highest return with the aim being to strengthen reward behavior and weaken punishment behavior (Zhang & Lu, 2021). This method mimics the way the humans and animals learn; therefore, there is no need to direct the agents' learning in a particular direction. Some common reinforcement learning algorithms include Q-learning, SARSA, and deep reinforcement learning.

1.3.3 National language processing

NLP is a field of study that combines computer science, linguistics, and artificial intelligence to enable computers to comprehend, translate, and generate human language. NLP is a crucial aspect of AI because language is a fundamental means of communication, and enabling computers to understand and use language has many practical applications (Baclic et al., 2020).

NLP comprises several subfields that work together to achieve various tasks related to language understanding and generation. These subfields include grammatical and semantic analysis, which involves identifying the structure and meaning of language; information extraction, which involves identifying and extracting relevant information from unstructured text data; text mining, which involves analyzing large amounts of text data to uncover patterns and trends; information retrieval, which involves finding relevant information from a large collection of text data; machine translation, which involves translating text from one language to another; question answering systems, which involve answering questions posed in natural language; and dialog systems, which involve enabling computers to interact with humans through natural language. Other practical applications of NLP include sentiment analysis, chatbots, speech recognition, and voice assistants. It has many potential applications in various other fields, including healthcare, finance, and customer service.

Natural language understanding (NLU) is the use of natural language for computer communication. The primary goal of NLU, also known as computational linguistics and located at the intersection of AI and language information processing, is to let computers understand the natural language. It captures sound signals as a machine, which NLP technology subsequently transforms into text signals and their meanings. The device then converts the sounds into words and the letters with meanings. These final two actions enable the machine to hear and comprehend. The system continuously learns to enhance the algorithm because it is equipped with speech recognition and semantic understanding technologies. As a result, in addition to hearing and interpreting, the system is also capable of recognizing emotions (Zhang & Lu, 2021).

1.3.4 Algorithm

Methods of abstraction have significant limits and limited precision. Researchers were inspired by the way babies learn, where no one teaches them how to recognize items but they manage to do so. As a result, machine learning strategies for summarizing the rules and methods for object recognition have been presented (Anantrasirichai & Bull, 2022). Given a large number of pictures of any living/non-living object, for example, a cat, the computer can study the qualities of cats using a variety of training models, such as neural networks, and then exactly identify cats in new pictures on the basis of these attributes.

The use of algorithms is critical in the field of artificial intelligence, allowing computers to automatically analyze and learn from enormous information and then make informed conclusions and predictions about real-world events based on the analysis. Beyond pattern recognition, machine learning algorithms have also achieved remarkable results in various other fields, including speech recognition, semantic analysis, recommendation systems, and search engines. These advancements have been largely driven by the development and implementation of AI algorithms.

1.3.5 Hardware

To solve complicated issues, machine learning makes use of some "deep" neural network models (Janiesch et al., 2021). Machine learning techniques like deep learning are used. NVIDIA GPUs are the primary hardware platform for deep learning operations. A modern computer strategy called GPU acceleration makes use of extremely parallel processors to speed up programs with parallel functions. Obtaining the training results used to take a CPU up to a month but currently, the GPU can do the same in a day. The GPU's powerful parallel computing capabilities reduce the deep learning algorithms' training time bottleneck, which unleashes the full potential of AI.

1.3.6 Computer vision

Computer vision is an area of research aimed at enabling computers to see and know the world through visual information, similar to how people do. It involves using algorithms for identifying images and analyzing them, with facial recognition and image recognition being two of the most commonly used computer vision applications (Olveres, 2021).

Deep learning has become widely used for image classification since 2015, utilizing neural networks with nonlinear fitting functions through activation functions. The model can learn feature extraction and categorization automatically given the correct input and output. Deep learning is used to streamline an intricate and time-consuming picture classification process, increasing its effectiveness and efficiency. Common neural network structures used in engineering include VGG, ResNet, and Startup, with Faster R-CNN, Mask-RCNN, and YOLO being the most commonly used network models for real-time facial recognition, which require both high accuracy and speed.

Image processing via computer vision is often done pixel by pixel, with semantic segmentation used to establish the meaning of the segmented pixels. Semantic segmentation involves distinguishing between dense pixels to recognize objects such as people, cars, motorcycles, and streetlights in an image. Convolutional neural networks have been used to develop semantic segmentation algorithms, replacing the fully linked layer of the network with fully convolutional networks (FCNs). FCN uses an encoder-decoder architecture that, after the encoder harvests features, uses a decoder to restore space size and specific information. To increase segmentation accuracy, a further technique called conditional random field (CRF) is created.

1.4 AI RESEARCH AREAS

There are several research areas where AI can be used. A few of them are depicted in Figure 1.3 and discussed below.

1.4.1 Expert system

Expert system is a form of AI that utilizes the knowledge and experience of human experts to solve complex problems in specific fields (Confalonieri et al., 2021). It is one of the earliest AI research fields and has found extensive applications in various industries such as geological surveying, medical diagnosis, and petrochemicals. The expert system uses professional knowledge, reasoning, and analysis to solve complicated problems that traditionally need an expert in the field in order to imitate the thought process of domain experts. The system has a vast database of information and reasoning processes, including professional knowledge and experience,

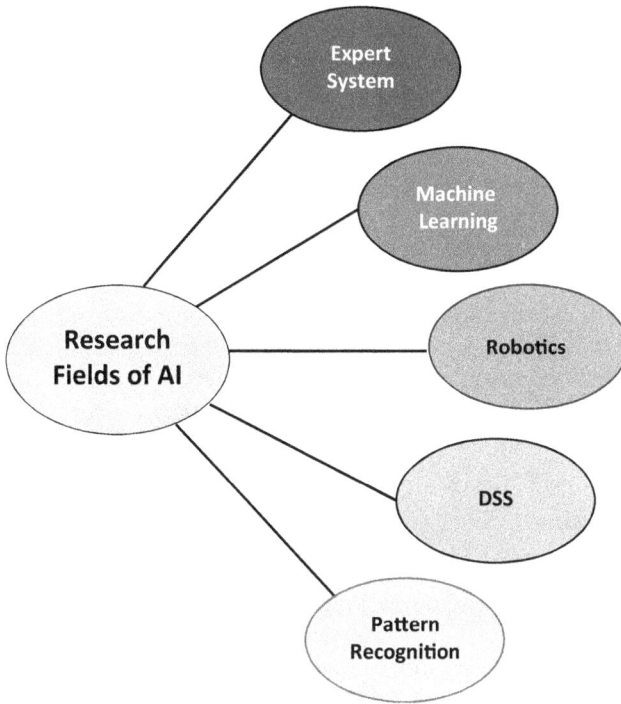

Figure 1.3 Research areas of artificial intelligence.

that can be accessed, stored, and judged simultaneously. The two essential components of an expert system are the knowledge base and reasoning engine (Zhang & Lu, 2021).

In order to use an expert system, the information, expertise, and research data of specialists in a given topic are first saved in a database and knowledge base. Then, as needed, the interpreter and reasoning engine call on this knowledge to provide solutions to problems through a computer-human interaction interface. Expert systems have advantages in teaching, as they are not limited by time, space, environment, or emotional influences. They can be used widely in education, particularly in distance learning, where their advantages are well known.

1.4.2 Machine learning

In order for a computer to possess information, the information must be formatted in a way that the computer can understand, or the computer must be capable of acquiring and continually enhancing knowledge on its own. This capability is referred to as machine learning (Janiesch et al., 2021). Machine learning research is primarily focused on three objectives: (1) studying human learning mechanisms and cognitive processes, (2) examining how people learn, and (3) creating learning systems for specific tasks. Machine learning research draws from various fields, such as medicine, information science, fuzzy mathematics, logic, neuropsychology, and brain science.

The deep learning concept is based on artificial neural networks. Restricted Boltzmann Machine (RBN), convolutional neural network (CNN), deep belief network (DBN), and stacking auto-encoder (Litjens et al., 2017) are included in deep learning algorithms. The traditional backpropagation algorithm had difficulty training neural networks with more than four layers. A deep learning structure comprises of multilayer perceptron with several hidden layers. By creating high-level attribute categories or features from low-level attributes, deep learning reveals the properties of data distribution. Important artificial neural network techniques include learning vector quantization (LVQ), self-organizing map (SOM), self-organizing network (SON), and perception neural network (PNN) (Zhang & Lu, 2021).

1.4.3 Robotics

To simulate human behavior, a machine is designed and called a robot (Karnouskos, 2022). The development of robots has gone through three generations of evolution. The first-generation robots are programmed by their designers, and the robot executes the program step by step (Boddu et al., 2022). The operations that the robot performs are recorded as instructions on the ground. The second generation of adaptive robots, however, have sensory sensors like touch, hearing, and vision that can gather basic data about their surroundings and the objects they are manipulating. The robot's actions are controlled by a computer. The third-generation robots, often referred to as intelligent robots, are outfitted with extremely sensitive sensors and exhibit intelligence akin to that of humans (Kim et al., 2004). The robot has the ability to analyze data, react to environmental changes, govern its behavior, and carry out difficult tasks.

1.4.4 Decision support system

The decision support system is closely related to the field of management science and is connected to the concept of "knowledge-intelligence." The expert system achieved great success in the 1980s, particularly in the field of AI, where intelligence and knowledge processing techniques have extended the application of decision support systems and enhanced their problem-solving capabilities. As a result, decision support systems have evolved into intelligent decision support systems (Eom & Kim, 2006).

1.4.5 Pattern recognition

The field of pattern recognition is concerned with equipping machines with the ability to perceive and identify visual and auditory patterns such as images, fonts, terrain, and objects. This technology is widely applicable in various aspects of daily life and military applications (Liu et al., 2006). The traditional statistical models and structural pattern recognition techniques are being gradually replaced by the rapidly evolving artificial neural network models and fuzzy mathematical models.

1.5 APPLICATIONS OF AI

AI has started to provide practical solutions and generate economic benefits by utilizing advanced technologies such as data processing, computing power, and algorithm. Finance, healthcare, automotive, and retail are some examples of sectors with solid data foundations that have sophisticated AI applications in their respective fields (Gupta et al., 2021). Some of the AI applications are depicted in Figure 1.4.

1.5.1 Automotive industry

Taking the example of autonomous driving in the automotive industry, this is a result of the amalgamation of cutting-edge technologies such as Internet of Things, AI, and the automotive industry (Tubaro & Casilli, 2019). It represents a significant development in global transportation and travel, as it uses advanced sensors like lidar and other devices to collect road and pedestrian data. These data are then combined with sophisticated AI algorithms to optimize the finest route and control strategy for self-driving cars on the road.

China has made significant strides in the field of autonomous driving and is quickly closing the gap with the United States and Europe,

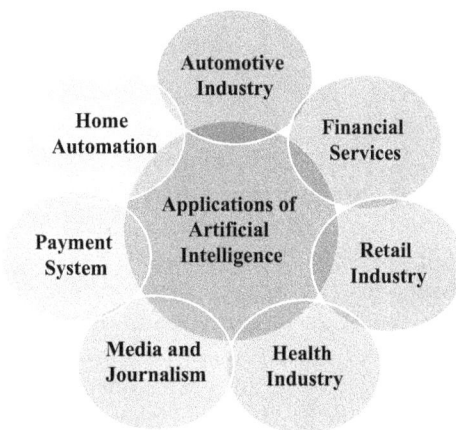

Figure 1.4 Applications of artificial intelligence.

attaining simultaneous development. Google made an announcement in December 2014 about the development of the first prototype of a driverless car. At the same time, Switzerland and France teamed up to manufacture driverless buses. In 2017, German car manufacturer Audi introduced its trademark "Audi AI" and used the technology of AI in the Olympic Games. Fixed-speed cruise along with automatic car's parking and other auxiliary drive devices enabled it to achieve a certain level of autonomous driving, thus reducing the need for manual driving tasks and liberating drivers.

1.5.2 Financial market

The financial markets have effectively incorporated AI technology in a number of sectors, including credit rating, market forecasting, consulting, and intelligent risk control, among others (Enholm et al., 2022). AI integration has ushered in a new era of financial innovation, with leading firms in the Silicon Valley using AI algorithms to increase user accessibility to financial goods. These models are trained with the financial analyst knowledge to track clients and reduce expenses. Users can quickly locate foreign exchange trading charts among a sea of data thanks to a deep learning algorithm developed by the Japanese company Alpaca, which scans and recognizes images. The AI market in the financial sector is expanding as machine learning is used to forecast risks and the direction of the stock market (Chhajer et al., 2022). In order to manage financial risks, combine various data sources, and give consumers real-time risk warnings, the financial institutions are implementing machine learning. Big data is also being utilized to analyze financial risks, offer in-the-moment risk warnings, conserve investment and financial management resources, create an academic risk management system, and lay the groundwork for business expansion.

1.5.3 Retail industry

AI is being used in the retail industry to enable unmanned physical stores, resulting in cost reduction and improved efficiency. Amazon, the e-commerce giant, has established a smart physical retail store called AmazonGo, which has quickly promoted smart retail (Türegün, 2019). The "Just Walk Out" system from AmazonGo tracks the addition and deletion of items from virtual shopping carts as well as from shelves using machine learning, sensors, and computer vision. E-commerce websites may boost online sales, produce more precise market forecasts, and spend less on inventory by incorporating AI into recommendation systems. On the basis of prospective customer preferences, online product recommendation models have been developed using the recommendation system.

1.5.4 Health industry

AI-based algorithms are extensively utilized in the medical field for various purposes, such as medical assistance, cancer detection, and drug development (Gupta et al., 2021; Olveres, 2021). The global promotion of medical information is crucial for the advancement of medical practices, and AI plays a significant role in achieving this objective. One of the most notable examples of AI in healthcare is IBM's intelligent robot, Watson. Initially, a significant amount of data and information, including medical data and reports, clinical recommendations, drug use reports, and hundreds of patient medical records, were entered into Watson by the IBM technical team. AI algorithms were then used to process and analyze the data, enabling stakeholders to make medical diagnosis and provide medical support more effectively and accurately.

1.5.5 Media and journalism industry

The media industry is implementing AI technology to enhance its ability to create content and communicate brand messaging (Trattner, 2022). One such implementation is the use of brand communication and content communication robots that can generate desired content with a single click and publish up to 10,000 articles per minute (Zhang & Lu, 2021). These intelligent media platforms utilize AI algorithms to analyze current events, public opinion, and PR marketing content to generate content that is relevant to users and follows media delivery rules. They can also automatically connect with mainstream media platforms to ensure effective dissemination. Additionally, to maximize communication value for the business, these platforms evaluate and match media items and channels based on brand content, marketing budgets, and anticipated results (Bailer et al., 2022).

1.5.6 Payment system

Currently, it has become common for people to shop cashless. Instead, they can effortlessly make payments using their mobile phones. If shoppers have not used the scan code payment method, there are new payment methods such as payment through voiceprint and face scanning. Shoppers can now go out without carrying their wallets, scanning a QR code, or entering a password. Voiceprint recognition is a recent technology that allows for the identification of people by their distinctive biological traits. For instance, Alipay Pay is a payment system based on AI, used in various restaurants, which can quickly distinguish between identical twins within a few seconds without any mistakes using the eye brushing technique. Brushing one's face has a varied range of applications, including eating, driving, and lodging. AI can also be helpful in detection of payment frauds (Soviany, 2018).

1.5.7 Home automation

Advanced technology is used in smart homes to integrate daily life facilities and manage family affairs to create comfortable living conditions. Smart houses give users smart intercommunication and include a variety of domestic devices like a TV, refrigerator, bathroom, door lock, air conditioner, etc. A complete smart home system consists of a number of items with various capabilities rather than simply one single device. Creating a self-learning system that can connect to other systems and learn new things about itself is the aim of home automation systems, which aim to effectively and intelligently mix household products and people.

Voice-controlled assistants have become a more convenient alternative to traditional interactive devices, such as buttons and touch screens. With the rise of smart homes, voice control has become a crucial entry point. To capitalize on this trend, major internet and technology companies have entered the market for smart speakers. In the realm of AI, smart speakers are viewed as the entry point for managing smart homes. Their primary function is to interact with humans through voice commands, and they have the ability to understand users' needs and provide appropriate services (Zhang & Lu, 2021).

1.6 THREE VIEWPOINTS OF AI

The field of AI research revolves around three main viewpoints of concepts, namely symbolism, connectionism, and behaviorism. These viewpoints are considered as the primary theoretical theorems that have played a crucial role in the development of AI disciplines, serving as their fundamental basis.

1.6.1 Symbolism

Symbolism, also known as logicism, asserts that symbols are vital to human cognition and that thinking is fundamentally a process of symbolic computation and reasoning (Zhang et al., 2023). Symbolism is used by first representing cognitive objects as symbols using mathematical logic, and then simulating human cognitive processes using a computer's symbol processing ability. Symbolism is founded on a physical sign system and the concept of limited reason. The technology of knowledge representation and logical reasoning lies at the heart of symbolism study. Symbolism research achievements include expert systems, knowledge engineering theory and technology, analytical reasoning approaches, and heuristic algorithms.

1.6.2 Connectionism

Connectionism, also known as bionics, asserts that human intelligence is intimately related to the physiological structure and function of the human brain. The core assumption of this method is that brain neurons are the constructing elements of human cognition, and cognitive processing is essentially the brain's processing of information. Thus, connectionism stresses the importance of mimicking the structure and functioning of the human brain to truly simulate human intelligence in machines. Connectionism study is mostly concerned with neural networks. Among the most notable achievements in this discipline are brain modeling studies and the creation of the backpropagation algorithm in multi-layer networks (Tian et al., 2022).

1.6.3 Behaviorism

Behaviorism, also known as the evolution of cybernetics, is based on the idea that action and perception—rather than knowledge, representation, or reasoning—define intelligence the most (Tilak et al., 2022). The ability to act, perceive, and sustain life and self-replication, according to behaviorism, is the most fundamental human ability. Interaction with the real world results in the production of intelligent behavior; thus, AI development should be progressive, just as human intelligence is. The primary purpose of behaviorism is to emulate various human control behaviors. The achievement of intelligent control and robot systems is a result of behaviorist progress. Although behaviorism has not yet evolved into a coherent theoretical theory, the AI community has paid close attention to it due to how it departs from traditional AI viewpoints.

These three perspectives have had a substantial and extensive influence on the development of artificial intelligence. They have fostered the emergence of a variety of particular AI analytical models, implementation strategies, and algorithms, and even the most modern knowledge discovery, data mining, and intelligent agent technologies are significantly impacted by these lookouts.

REFERENCES

Anantrasirichai, N., & Bull, D. (2022). Artificial intelligence in the creative industries: A review. Artificial Intelligence Review, 1–68.

Baclic, O., Tunis, M., Young, K., Doan, C., Swerdfeger, H., & Schonfeld, J. (2020). Artificial intelligence in public health: Challenges and opportunities for public health made possible by advances in natural language processing. Canada Communicable Disease Report, 46(6), 161.

Bailer, W., Thallinger, G., Krawarik, V., Schell, K., & Ertelthalner, V. (2022, March). AI for the media industry: Application potential and automation levels. In International Conference on Multimedia Modeling (pp. 109–118). Cham: Springer International Publishing.

Beam, A. L., Drazen, J. M., Kohane, I. S., Leong, T. Y., Manrai, A. K., & Rubin, E. J. (2023). Artificial intelligence in medicine. New England Journal of Medicine, 388(13), 1220–1221.

Boddu, R. S. K., Santoki, A. A., Khurana, S., Koli, P. V., Rai, R., & Agrawal, A. (2022). An analysis to understand the role of machine learning, robotics and artificial intelligence in digital marketing. Materials Today: Proceedings, 56, 2288–2292.

Chen, G., Huang, B., Chen, X., Ge, L., Radenkovic, M., & Ma, Y. (2022). Deep blue AI: A new bridge from data to knowledge for the ocean science. Deep Sea Research Part I: Oceanographic Research, 190, 103886.

Chhajer, P., Shah, M., & Kshirsagar, A. (2022). The applications of artificial neural networks, support vector machines, and long–short term memory for stock market prediction. Decision Analytics Journal, 2, 100015.

Confalonieri, R., Coba, L., Wagner, B., & Besold, T. R. (2021). A historical perspective of explainable artificial intelligence. Wiley Interdisciplinary Reviews: Data Mining and Knowledge Discovery, 11(1), e1391.

Enholm, I. M., Papagiannidis, E., Mikalef, P., & Krogstie, J. (2022). Artificial intelligence and business value: A literature review. Information Systems Frontiers, 24(5), 1709–1734.

Eom, S., & Kim, E. (2006). A survey of decision support system applications (1995–2001). Journal of the Operational Research Society, 57, 1264–1278.

Ghazal, T. M. (2021). Performances of K-means clustering algorithm with different distance metrics. Intelligent Automation & Soft Computing, 30(2), 735–742.

Gupta, R., Srivastava, D., Sahu, M., Tiwari, S., Ambasta, R. K., & Kumar, P. (2021). Artificial intelligence to deep learning: Machine intelligence approach for drug discovery. Molecular Diversity, 25, 1315–1360.

Huynh-The, T., Pham, Q. V., Pham, X. Q., Nguyen, T. T., Han, Z., & Kim, D. S. (2023). Artificial intelligence for the metaverse: A survey. Engineering Applications of Artificial Intelligence, 117, 105581.

Janiesch, C., Zschech, P., & Heinrich, K. (2021). Machine learning and deep learning. Electronic Markets, 31(3), 685–695.

Karnouskos, S. (2022). Symbiosis with artificial intelligence via the prism of law, robots, and society. Artificial Intelligence and Law, 30(1), 93–115.

Kim, J. H., Kim, Y. D., & Lee, K. H. (2004, December). The third generation of robotics: Ubiquitous robot. In Proceedings of the Second International Conference on Autonomous Robots and Agents.

Kosinski, W., & Zaczek-Chrzanowska, D. (2007). Pavlovian, Skinner, and other behaviourists' contributions to AI. Intelligent Motion and Interaction Within Virtual Environments.

Litjens, G., Kooi, T., Bejnordi, B. E., Setio, A. A. A., Ciompi, F., Ghafoorian, M., Van Der Laak, J. A., Van Ginneken, B., & Sánchez, C. I. (2017). A survey on deep learning in medical image analysis. Medical Image Analysis, 42, 60–88.

Liu, J., Sun, J., & Wang, S. (2006). Pattern recognition: An overview. IJCSNS International Journal of Computer Science and Network Security, 6(6), 57–61.

Olveres, J., González, G., Torres, F., Moreno-Tagle, J. C., Carbajal-Degante, E., Valencia-Rodríguez, A., Méndez-Sánchez, N., & Escalante-Ramírez, B. (2021). What is new in computer vision and artificial intelligence in medical image analysis applications. Quantitative Imaging in Medicine and Surgery, 11(8), 3830.

Owsley, C. S., & Greenwood, K. (2022). Awareness and perception of artificial intelligence operationalized integration in news media industry and society. AI & SOCIETY, 1–15.

Reddy, B., & Fields, R. (2022, April). From past to present: A comprehensive technical review of rule-based expert systems from 1980-2021. In Proceedings of the 2022 ACM Southeast Conference (pp. 167–172).

Saranya, T., Sridevi, S., Deisy, C., Chung, T. D., & Khan, M. A. (2020). Performance analysis of machine learning algorithms in intrusion detection system: A review. Procedia Computer Science, 171, 1251–1260.

Soviany, C. (2018). The benefits of using artificial intelligence in payment fraud detection: A case study. Journal of Payments Strategy & Systems, 12(2), 102–110.

Tian, S., Zhang, J., Shu, X., Chen, L., Niu, X., & Wang, Y. (2022). A novel evaluation strategy to artificial neural network model based on bionics. Journal of Bionic Engineering, 1–16.

Tilak, S., Glassman, M., Kuznetcova, I., & Pelfrey, G. L. (2022). Applications of cybernetics to psychological theory: Historical and conceptual explorations. Theory & Psychology, 32(2), 298–325.

Trattner, C., Jannach, D., Motta, E., Costera Meijer, I., Diakopoulos, N., Elahi, M., Opdahl, A. L., Tessem, B., Borch, N., Fjeld, M., & Øvrelid, L. (2022). Responsible media technology and AI: Challenges and research directions. AI and Ethics, 2(4), 585–594.

Tubaro, P., & Casilli, A. A. (2019). Micro-work, artificial intelligence and the automotive industry. Journal of Industrial and Business Economics, 46, 333–345.

Türegün, N. (2019). Impact of technology in financial reporting: The case of Amazon Go. Journal of Corporate Accounting & Finance, 30(3), 90–95.

Xu, L. D., Lu, Y., & Li, L. (2021). Embedding blockchain technology into IoT for security: A survey. IEEE Internet of Things Journal, 8(13), 10452–10473.

Zhang, B., Zhu, J., & Su, H. (2023). Toward the third generation artificial intelligence. Science China Information Sciences, 66(2), 121101.

Zhang, C., & Lu, Y. (2021). Study on artificial intelligence: The state of the art and future prospects. Journal of Industrial Information Integration, 23, 100224.

Zhang, C.-M., & Chu, H.-M. (2020, December). Preprocessing method of structured big data in human resource archives database. In 2020 IEEE International Conference on Industrial Application of Artificial Intelligence (IAAI) (pp. 379–384). IEEE.

Zhang, S. (2022). Research on energy-saving packaging design based on artificial intelligence. Energy Reports, 8, 480–489.

Zhang, Z., Ning, H., Shi, F., Farha, F., Xu, Y., Xu, J., ... & Choo, K. K. R. (2022). Artificial intelligence in cyber security: Research advances, challenges, and opportunities. Artificial Intelligence Review, 1–25.

Chapter 2

Building an inclusive future
Supportive environments for individuals on the autism spectrum

Rolly Gupta and Lalit Kumar Sagar
Department of Computer Science Engineering, SRMIST, Ghaziabad,
Uttar Pradesh, India

2.1 BACKGROUND

This is a component of a research that attempts to create a framework that enables people with autism to lead more autonomous lives. By creating and implementing an independent living platform in a smart society, the research project will achieve its stated goal of assisting people with autism in their struggles with various inclusion challenges. Making sure that this platform is adaptable to user needs is one of the hurdles in properly deploying it. Investigating how this might be done successfully is thus worthwhile.

According to the creator's predetermined rules, players in the metaverse exist in a virtual universe (Hwang & Chien 2022). The "next big thing" in technology is how people commonly describe it. However, the educational potential of the metaverse is generally overlooked (ibid). This is a rising use of the metaverse and is referred to as "immersive learning" (De Freitas et al. 2010). Currently, military and medical training are two fields where immersive learning is applied (Herrington et al. 2007).

Due to their difficulty planning the steps necessary to complete a task and their tendency to pay less attention to significant environmental cues, people with autistic spectrum disorder (ASD) can benefit from this training (Hume et al. 2009). The availability of a secure, virtual environment for autonomous creation where these challenges can be taken into account is underutilized in current research and practice.

The ability of such a virtual environment to adapt to the user is essential to its success. The user's intrinsic characteristics can be captured by machine learning in addition to these abstract criteria. El Aissaoui et al. (2019) and Chen et al. (2020) found that this approach works well in educational settings. While there are now research projects designed to help ASD people participate in the virtual world, these are generally participation-based and do not concentrate on enhancing abilities for

DOI: 10.1201/9781003461418-2

independent living (Bian et al. 2019). A category of accessible metaverses that aid a larger range of users in their daily lives can thus be fostered using the ability of machine learning to replicate complex user behavior.

2.2 PROBLEM

The majority of significant metaverse implementations as of right now are not created to be accessible or flexible (Mott et al. 2019). This indicates that they are essentially unchangeable and don't change in a significant way in respect to the user. This is problematic since the metaverse has a lot of untapped potential if 3D-rendered virtual worlds' inherent adaptability is not utilized (Hwang & Chien 2022).

For groups with particular needs, such as individuals with ASD, these "fixed" types of metaverses could erect a barrier to entry that is challenging to get over. A non-adaptive metaverse might not provide the clear structure and unambiguous information that people with ASD need (Lord et al. 2018). For different groups of individuals to use the metaverse, this issue needs to be resolved. Investigating the same approach for underprivileged populations should be of relevance given the growing usage of the metaverse as an educational platform (Herrington et al. 2007). A very inclusive space that can accommodate people with different types of ASD and give them access to the metaverse as a learning environment has a ton of promise.

The problem is that current mainstream metaverse implementations are neither accessible or adaptable, which puts people with ASD who need straightforward structure and information at a disadvantage. Because of this, they struggle to fully utilize the metaverse as a platform for inclusive education.

2.3 GOAL

The goal of this effort is to develop an artifact that can adapt an interactive environment to a user's needs using machine learning. Since people with ASD are the core of the identified problem, these user criteria will be tailored exclusively to them. Additionally, this research is expected to advance the body of knowledge regarding the usage of virtual worlds in a variety of educational settings. This results in the research topic that follows, which is a direct result of the mentioned issue:

> "How can the current virtual reality platforms be linked with machine learning-based 3D environment adaptations to increase accessibility for ASD users?"

2.4 DELIMITATIONS

The main goal of this research is to use machine learning to adapt 3D environments for individuals with ASD. As a result, the larger question of accessibility in metaverse design or in the situation of other excluded groups will not be covered in this study. This effort will focus on machine learning–based virtual environment adaptability rather than the PSP (Stockholm University 2022) platform's development. The study will not consider any other potential applications and will be restricted to the educational setting of the metaverse. This work uses simulated data due to time and cost limitations, the effects of which will be fully discussed.

2.5 METHODOLOGY

It is intriguing to see how the process of creation is described because both the issue statement and the research question of this work require the construction of an artifact. It is crucial to apply the appropriate framework to evaluate these potential benefits because the main objective of the suggested artifact is to determine how machine learning might help people with ASD. The definition of design science research by Johannesson and Perjons (2021) makes it a suitable framework for aiding in the creation and assessment of this kind. According to Johannesson and Perjons (2021), this type of design science research is categorized as an improvement, in which fresh approaches to long-standing issues are investigated. The innovative approaches in this paper make use of machine learning to address a well-known problem, namely accessibility for individuals with ASD. Four design science framework elements are established by Johannesson and Perjons (2021) to support study in:

- Dividing the work into logically connected tasks with clear inputs and outputs.
- Executing these tasks
- Deciding on research techniques and procedures to apply to activities
- Relating the results to the corpus of knowledge already in existence

There are a number of disadvantages to design science research that must be weighed against the advantages of using it as a research paradigm. Any artifacts produced might have poor generalizability, which would make it challenging to apply the research's solutions to other problems. Because the purpose of this study is to specifically evaluate a remedy for getting accurate result.

In a particular situation, this is challenging to resist and is covered by the restrictions of the work. The subjective nature of appraising one's own research may prevent objective evaluation, which is another issue with design science research. To get the most accurate study results possible, this will be considered while establishing the specifications for the artifact. Any ethical or societal ramifications of design science research that are pertinent to this work are another restriction.

According to Denscombe (2014), this effort will also employ experiment as a research methodology to make sure that an accurate assessment of the artifact is made. Consequently, the specific type of experiment will be a laboratory experiment in which the artifact and all of its variables will be observed under supervised circumstances. The repeatability of this method enhances the iterative nature of the creation process and improves the reliability of the finished product. Experiments are also quite precise, so any data gathered can be utilized to build a strong case for more extensive discussions. The impact of its artificial settings is one drawback of utilizing this strategy, though. Results from an ex ante evaluation that is carried out without the involvement of any actual stakeholders might not be applicable to real-world use cases.

2.6 DESIGN AND DEVELOPMENT

An initial design of the artifact's planned practical application is shown in Figure 2.1. In this scenario, either textual data about the user or information from a person involved in the user's daily activities is used to train a machine learning model. The virtual world is then modified by the 3D engine based on the user's specified attributes.

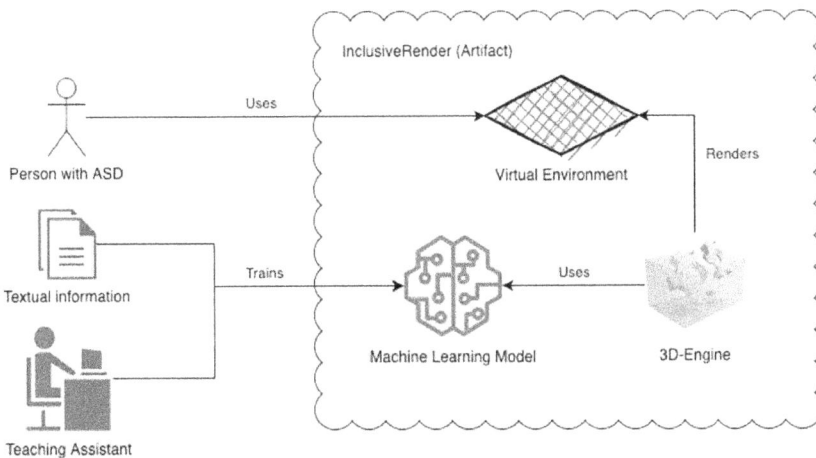

Figure 2.1 Preliminary system overview.

2.7 DISCUSSION

This offers a clear response to the research problem by successfully integrating machine learning into virtual reality platforms to modify virtual environments for people with ASD.

"How can existing virtual reality platforms be improved for individuals with ASDT by integrating machine learning-based 3D environment adaptations?"

The strategy is effective for the majority of unseen sorts of people (8/10) and handles known user profiles with 87.7% accuracy. This approximation, which is the main justification for utilizing machine learning as a solution, may result in even better results given a wider scope and the use of more varied data creation techniques. With its many virtual aids (pictograms, texts, and infographics), the artifact encourages immersive learning for individuals with ASD. It is simple to add new assistants; however, doing so requires interaction with the Unity interface, which was not separately developed. No significant issues with the artifact's effectiveness or performance have been raised, and neither its development nor review has uncovered any significant flaws.

2.8 SIGNIFICANCE

The benefits of using machine learning in education have been demonstrated by El Aissaoui et al. (2019). By using neural networks instead of the Naive-Bayes Classifier and K-modes, this work expands on that work.

Using clustering, El Aissaoui and associates (2019) conducted their study. Even though the aforementioned algorithms are appropriate for straightforward and brief cases, they have limits. Since it assumes independent qualities, the Naive-Bayes Classifier is inadequate for building complicated user models. The K-modes clustering algorithm, on the other hand, is intended for categorical data. This study shows how neural networks may be used to more accurately represent user groups and handle complicated interactions. Based on these intricate feature linkages, the use of neural networks enables more precise estimates of unknown user profiles. Neural networks do have some downsides, such as the need for huge labeled data sets and expensive computational expenditures, which may increase the time and resource requirements of any produced artifacts.

2.9 CONCLUSION

Through the successful integration of machine learning approaches, this work enables the modification of virtual environments to accommodate individuals with ASD. The system can approximate unknown user profiles while handling known user profiles with accuracy. People with ASD benefit from immersive learning experiences when different types of virtual assistants are used. To improve the reliability and validity of the findings, additional research is required, such as ethnographic studies and the collection of real data. With the development of generative AI technologies, opportunities for user profile analysis and context-aware virtual support have emerged. Future research involving actual data must take ethical issues like informed consent and privacy into account.

REFERENCES

Bian, D., Wade, J., Swanson, A., Weitlauf, A., Warren, Z. & Sarkar, N. (2019), 'Design of a physiology-based adaptive virtual reality driving platform for individuals with asd', *ACM Transactions on Computer Systems* 12(1), 1–24.

Chen, L., Chen, P. & Lin, Z. (2020), 'Artificial intelligence in education: A review', *IEEE Access* 8, 75264–75278.

De Freitas, S., Rebolledo-Mendez, G., Liarokapis, F., Magoulas, G. & Poulo vassilis, A. (2010), 'Learning as immersive experiences: Using the four dimensional framework for designing and evaluating immersive learning experiences in a virtual world', *British Journal of Educational Technology* 41 (1), 69–85.

Denscombe, M. (2014), *The good research guide: For small-scale social research projects*, 5 edn, Open University Press, Buckingham, England.

El Aissaoui, O., El Madani, Y. E. A., Oughdir, L. & El Allioui, Y. (2019), 'Combining supervised and unsupervised machine learning algorithms *to* predict the learners' learning styles', *Procedia Computer Science* 148, 87–96.

Herrington, J., Reeves, T. C. & Oliver, R. (2007), 'Immersive learning technologies: Realism and online authentic learning', *Journal of Computing in Higher Education* 19, 80–99.

Hume, K., Loftin, R. & Lantz, J. (2009), 'Increasing independence in autism spectrum disorders: A review of three focused interventions', *Journal of Autism and* Developmental Disorders 39, 1329–1338.

Hwang, G.-.J. & Chien, S.-Y. (2022), 'Definition, roles, and potential research issues of the metaverse in education: An artificial intelligence perspective', *Computers and Education: Artificial Intelligence* 3, 100082.

Johannesson, P. & Perjons, E. (2021), *An introduction to design science an introduction to design science*, 2 edn, Springer Nature, Cham, Switzerland.

Lord, C., Elsabbagh, M., Baird, G. & Veenstra-Vanderweele, J. (2018), 'Autism spectrum disorder', *The Lancet* 392(10146), 508–520.

Mott, M., Cutrell, E., Franco, M. G., Holz, C., Ofek, E., Stoakley, R. & Morris, M. R. (2019), 'Accessible by design: An opportunity for virtual reality', in 2019 IEEE International Symposium on Mixed and Augmented Reality Adjunct (!SMAR-Adjunct), IEEE, pp. 451–454.

Stockholm University (2022), 'Platform for smart people', Accessed: 2023-05- 24.

Chapter 3

Cloud computing and virtualization

A symbiotic revolution in modern IT infrastructure

Jaishree Jain and Ashish Dixit

Department of Computer Science & Engineering, AKGEC, Ghaziabad, Uttar Pradesh, India

3.1 INTRODUCTION

Virtualization and cloud computing are two different but related ideas in the field of information technology. "Cloud computing" is the term used to describe the pay-per-use delivery of computer services, such as storage, processing power, and software applications, over the internet. Customers can remotely access these resources through the internet from cloud service providers' data centers. Customers may instantly scale up or down their computer capabilities as needed without investing in expensive hardware and infrastructure thanks to the scalability, flexibility, and cost-effectiveness of cloud computing.

3.2 DEFINING CLOUD COMPUTING

Cloud service providers, who own and manage the infrastructure needed to deliver the services, offer these services. Some of the most popular cloud computing services are listed below.

- *Infrastructure as a Service (IaaS):*
  This service gives users online access to virtualized computer resources like networking, storage, and virtual machines. On top of the virtualized infrastructure offered by the cloud service provider, customers can deploy and manage their own applications.
- *Platform as a Service (PaaS):*
  This service gives clients access to a platform so they may create, distribute, and maintain their own apps without having to worry about the infrastructure supporting them. Typically, PaaS services include tools and frameworks that aid programmers in creating and testing applications.
- *Software as a Service (SaaS):*
  This service offers online access to software programs without requiring the installation or upkeep of the programs on local

DOI: 10.1201/9781003461418-3

computers. Web browsers or specific programs can be used to access SaaS applications.

- *Database as a Service (DBaaS):*
  This service provides access to managed database services, such as MySQL, PostgreSQL, and MongoDB, over the internet. DBaaS services typically provide features such as automated backups, scaling, and monitoring.
- *Function as a Service (FaaS):*
  This service provides a platform for customers to deploy and run serverless applications, which are event-driven and automatically scale based on demand. FaaS services typically charge customers based on the number of function invocations and the duration of each invocation [1].
- *Content Delivery Network (CDN):*
  This service provides a globally distributed network of servers that can cache and deliver static and dynamic content, such as images, videos, and web pages, to users based on their geographic location. CDNs can improve the performance and reliability of web applications by reducing the latency and network congestion.
- *Serverless Computing:*
  With serverless computing, developers only pay for the computing resources used when their code runs, making it a cost-effective option (Figure 3.1).

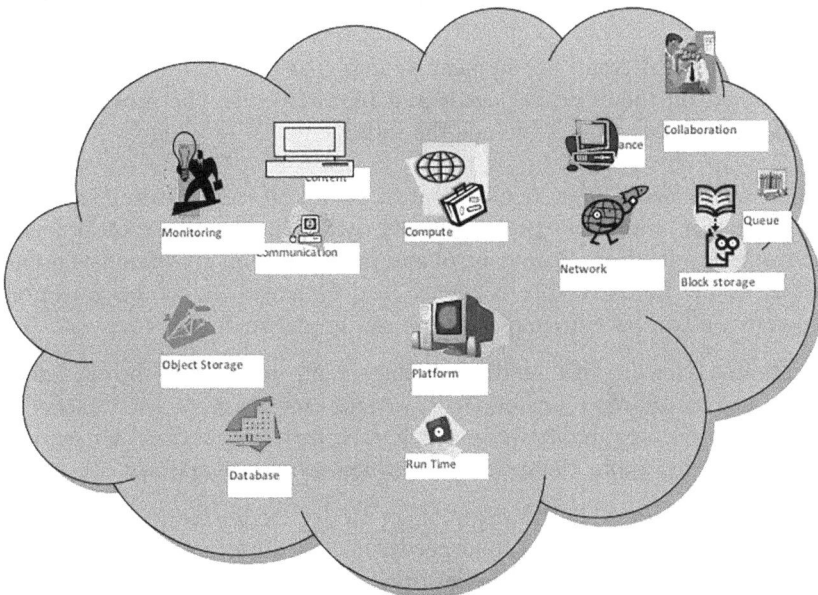

Figure 3.1 Cloud computing.

- *Cloud storage:*
  Cloud storage provides on-demand access to data storage over the internet. This allows organizations to store and retrieve large amounts of data without needing to manage their own storage infrastructure.
- *Cloud security:*
  Cloud security includes a range of tools and processes designed to protect cloud resources from unauthorized access, data breaches, and other security threats. It typically includes tools such as firewalls, encryption, and access controls. These components work together to provide a flexible, scalable, and cost-effective computing environment for organizations of all sizes.
- *Multi-cloud:*
  It offers flexibility and avoids vendor lock-in, but can be more complex to manage. Each deployment model has its own advantages and disadvantages, and organizations should carefully consider their requirements before choosing the best option.

  Overall, cloud computing services provide a flexible and cost-effective way for customers to access and consume computing resources, without having to manage the underlying infrastructure. Cloud services can also provide scalability, reliability, and security features that are difficult to achieve with traditional on-premises computing infrastructure.

3.3 CLOUD DEPLOYMENT MODELS

There are several cloud deployment models that organizations can choose from based on their specific needs and requirements. The most common cloud deployment models include the following.

a. Private cloud: This deployment model involves creating a cloud infrastructure within an organization's own data center or on-premises environment. It offers more control and customization compared to public cloud, as the infrastructure is owned and managed by the organization. However, it can be more expensive to set up and maintain.

b. Hybrid cloud: This model combines both public and private cloud environments, allowing organizations to take advantage of the benefits of both. It enables organizations to maintain their own private cloud for sensitive data while using public cloud resources for less sensitive workloads [2].

c. Community cloud: In this deployment model, cloud resources are shared among organizations with similar needs, such as those in the same industry or geographical location. It enables organizations to share costs while still maintaining control over their data.

d. Multi-cloud: It offers flexibility and avoids vendor lock-in, but can be more complex to manage (Figure 3.2).

User

| Public Cloud | Private Cloud | Hybrid Cloud |

Cloud Service Provider

Infrastructure & Server Focus

User

| Web Browser | Public App | API |

Ethernet(Public Network)

| Load Balancer | Firewall | Router |

Cloud Service Provider

| Virtual Memor | Object Storage | Database |

Cloud Infrastructure

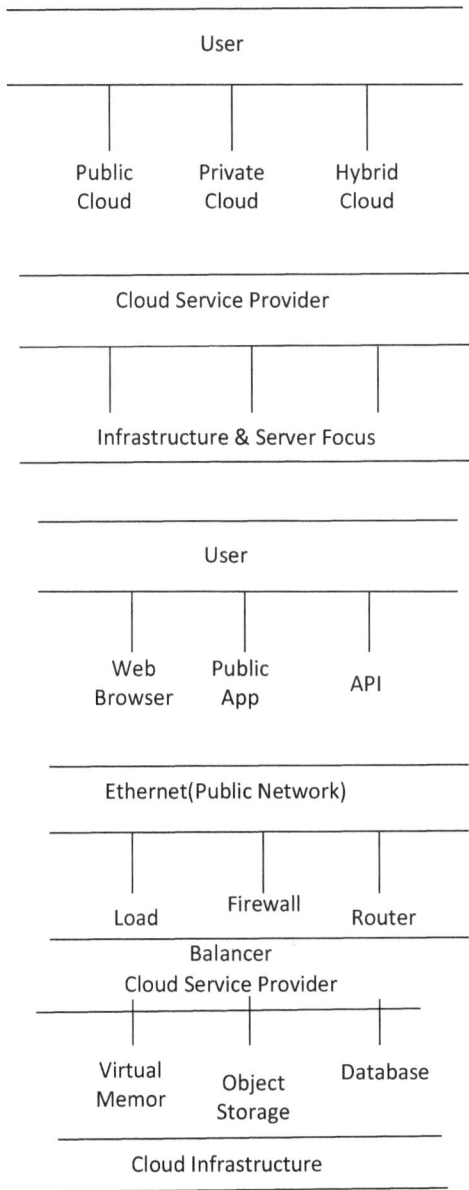

Figure 3.2 Architecture of cloud computing.

The user here has access to cloud resources via a public, private, or hybrid cloud. A third-party cloud service provider offers the public cloud, which is available to anyone online. The private cloud is exclusively used by one company and is typically hosted on-site or by an outside provider.

Organizations can use the advantages of both public and private clouds by combining them in the hybrid cloud.

Users can access virtual computers, storage, and databases using the cloud service provider's infrastructure and services. The level of customization and control an organization has over its infrastructure and services depends on the type of cloud being employed. For instance, compared to a private cloud, a public cloud often provides less control and personalization [3].

Each deployment model has its own advantages and disadvantages, and organizations should carefully consider their requirements before choosing the best option.

3.4 WORKINGS OF CLOUD COMPUTING

Cloud computing works on several key principles that enable organizations to use computing resources over the internet. The main principles of cloud computing include the following.

- *On-demand self-service:*
  Users can quickly and easily provision computing resources such as servers, storage, and networking on demand, without requiring human intervention from the service provider.
- *Broad network access:*
  Cloud services can be accessed from anywhere with an internet connection, using a range of devices such as laptops, smartphones, and tablets.
- *Resource pooling:*
  Cloud resources are shared among multiple users, allowing them to access computing resources as needed without requiring dedicated hardware.
- *Rapid elasticity:*
  Cloud resources can be quickly and easily scaled up or down to meet changing demands. This enables organizations to easily accommodate spikes in usage and avoid over-provisioning [4].
- *Measured service:*
  Cloud services are typically metered and billed based on usage, allowing organizations to only pay for what they use (Figure 3.3).
  In this diagram, the user accesses cloud resources through various interfaces such as web browsers, mobile apps, or APIs/SDKs. These interfaces communicate with the cloud service provider over the internet through load balancers, firewalls, and routers. The cloud service provider owns and manages the cloud infrastructure, which includes virtual machines, object storage, and databases. The infrastructure is typically composed of multiple data centers located in different geographical locations.

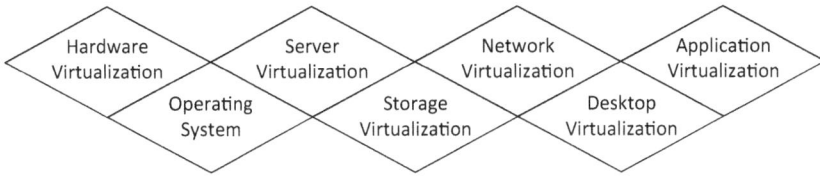

Figure 3.3 Types of cloud computing.

3.5 VIRTUALIZATION

A single physical server or computer may now run numerous operating systems or apps thanks to the technology known as virtualization. Making a virtual copy of a server, operating system, storage device, or network resource is the process of virtualization. Virtualization is frequently used in the context of computing to simulate several computing environments on a single physical computer and can run numerous operating systems or applications concurrently. This makes it possible to use hardware resources more effectively, lowers hardware costs, and makes it simpler to manage computer resources [5].

3.6 CLOUD COMPUTING AND VIRTUALIZATION

Virtualization technology is frequently used in cloud computing to allow the sharing of computing resources among many consumers. Cloud service providers can efficiently distribute resources and provide various levels of service to various customers by using virtualization to construct several virtual machines on a single physical server. Additionally, virtualization enables cloud service providers to build separate environments for each client, improving security and dependability [6].

Virtualization and cloud computing are complementary technologies that make it possible to distribute computing resources over the internet effectively and affordably. While virtualization facilitates the effective exploitation of those resources through the construction of virtual machines, cloud computing refers to the supply of computer resources [7].

Virtualization in cloud computing involves building virtual versions of resources including servers, storage, and networks. Software is then used to manage these virtual instances, enabling a more flexible and effective use of resources.

3.7 EQUIPMENT ABSTRACTION

A server, storage device, or network's physical hardware resources are abstracted by virtualization into manageable virtual instances.

- *Resource pooling:*
  Virtualization enables the sharing of resources across numerous physical devices, including CPU, memory, and storage. This makes it possible to use resources more effectively and cuts down on waste.
- *Hypervisor:*
  A virtual machine monitor (VMM), also referred to as a hypervisor, runs on top of the host operating system [8].
- *Containerization:*
  Applications can run in isolated environments called containers thanks to a lightweight virtualization technique called containerization. Containers share the host system's operating system kernel, which lowers virtualization costs and enhances performance.

  Cloud computing is based on several essential technologies, including virtualization. It enables the development of virtualized variations of computing resources like servers, storage, and networking that may be dynamically assigned and de-allocated as necessary. A program known as a hypervisor or virtual machine monitor (VMM) is used to accomplish this [9]. A physical server's hypervisor can support many virtual machines (VMs) running on top of it. Each VM is separate from the others and presents itself to the user as such a distinct physical device. The hypervisor controls how the VMs are allotted physical resources like CPU, memory, and storage. This makes it feasible to use physical resources considerably more efficiently than would be possible with just one dedicated server [10].

  In cloud computing, the infrastructure as a service (IaaS) layer is made available using virtualization. Pools of virtual resources can be dynamically allocated to consumers on demand by cloud service providers using hypervisors. Following that, clients might employ these resources to conduct their own programs and services.

  The advantages of virtualization in cloud computing include the following.
- *Increased resource usage:*
  As numerous VMs can run on a single physical server, virtualization enables more effective use of physical resources.
- *Scalability:*
  Applications and services can be scaled quickly due to the ability to dynamically allocate and de-allocate virtual resources as needed.
- *Flexibility:*
  Customers can choose the virtual resources' configuration and type, and they can quickly alter them as their needs change.
- *Cost savings:*
  Virtualization can considerably lower the cost of IT infrastructure by sharing physical resources.

3.8 TYPES OF VIRTUALIZATIONS IN CLOUD COMPUTING

Several types of virtualizations are used in cloud computing. These include the following.

- *Server virtualization:*
  This involves the creation of multiple virtual servers on a single physical server. Each virtual server operates independently and can run different operating systems and applications.
- *Network virtualization:*
  This allows multiple virtual networks to run on a single physical network. Each virtual network is isolated from the others and can have its own policies and security controls.
- *Storage virtualization:*
  This involves the abstraction of physical storage devices into virtual storage resources. This allows for more efficient use of storage capacity and enables features such as data deduplication, snapshotting, and replication.
- *Desktop virtualization:*
  This allows multiple virtual desktops to run on a single physical machine. Each virtual desktop operates independently and can be accessed remotely by end users.
- *Application virtualization:*
  This involves the encapsulation of applications into virtual packages that can be run on any operating system or hardware platform. This allows for greater flexibility and mobility of applications and can simplify software deployment and management [11].
- *Operating system virtualization:*
  This involves the creation of multiple virtual instances of an operating system on a single physical machine. Each instance operates independently and can run different applications (Figure 3.4).

Types of Virtualizations in Cloud Computing

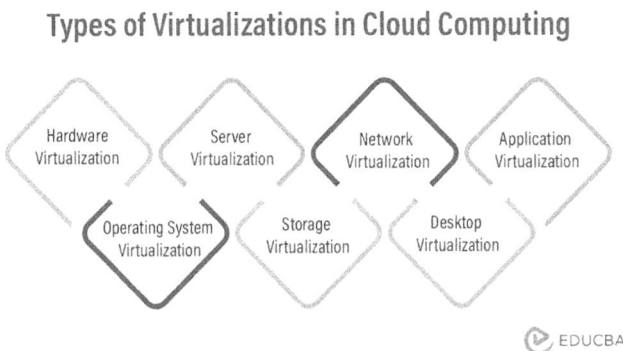

Figure 3.4 Types of virtualization techniques.

Each type of virtualization offers different benefits and is suited for different use cases. Cloud providers typically offer a combination of these virtualization types to provide a flexible and scalable cloud infrastructure [12].

3.9 ADVANTAGES

The following are benefits of virtualization in cloud computing:

- More Economical: Save money and the environment.
- Facilitating Agile: Promotes effective and adaptable operations.
- Easy to locate, transfer, and retrieve data.
- Efficient and flexible data transfer.
- Zero Chance of System Failure: Clustering guarantees ongoing operations.

3.10 LIMITATIONS

- Overhead: Extra resource needs that could affect performance.
- Complexity: Increases the complexity of managing, troubleshooting, and securing IT infrastructure.
- License: For virtualized environments, some software suppliers may charge extra license fees.
- Single Point of Failure: The availability of essential applications may be impacted by a physical server failure.
- Security: New security concerns, like VM escape attacks, call for further security precautions.

3.11 CLOUD COMPUTING WITH AMAZON

This chapter is a comprehensive general reference for developers and systems administrators who are creating transactional web applications in any language. But the bulk of cloud workers associate "cloud infrastructure" with Amazon EC2 and Amazon S3 [13]. This fact, coupled with the fact that I used examples from the Amazon cloud, necessitates a description of cloud computing generally, as well as the Amazon cloud.

3.12 S3 AMAZON

Persistent cloud storage is provided by Amazon Simple Storage Service (S3). Other Amazon services are not dependent on it to function. In fact, you may use Amazon S3 in apps you create for hosting on your own servers without ever actually "being in the cloud." The feature set, not the simplicity of use, is what Amazon means when it calls S3 "easy storage." You can easily upload data to the cloud and download it again thanks to Amazon S3. You do not need to be familiar with the method or location of storage [14].

If you consider Amazon S3 to be a remote file system, you are doing Amazon S3 a grave disservice. In many ways, Amazon S3 is considerably more basic than a file system. In reality, you actually store objects rather than "files." Additionally, rather than using directories, you store items in buckets. Despite the fact that these distinctions might only seem semantic, they actually reflect a number of significant variations:

The maximum size of an object saved in S3 is 5 GB [15].

- All Amazon S3 users share a flat namespace where buckets are located. There is no such thing as "sub-buckets," and namespace conflicts must be avoided.
- You can publish your buckets and items for public viewing.
- You cannot "mount" S3 storage without third-party software. In fact, I dislike using third-party tools to mount S3 since I think it is improper to treat it as a file system given how fundamentally distinct it is from one.

3.13 GETTING TO S3

You must register for an Amazon Web Services account in order to access S3. You can request default storage in either Europe or the United States. Your choice of data storage location is not just dependent on your residence. Regulatory and privacy issues will affect where you decide to keep your cloud data, as we shall cover later in this book. I advise using the storage unit that is nearest to the point of access to S3 for this chapter.

3.14 ONLINE SERVICES

Using a SOAP API and a REST API, Amazon [16]:

- Make fresh buckets
- Remove old buckets and objects
- Add new objects
- You can optionally define the place where the contents of your buckets should be kept while modifying them.
- But, you should absolutely download the s3cmd command-line client for Amazon S3 (http://s3tools.logix.cz/s3cmd) in order to get started with Amazon S3. The S3 access web services are wrapped in a command-line interface.

3.15 CLOUD SECURITY

If the cloud makes you completely reevaluate how you think about any aspect of your infrastructure in particular, it's probably security [17]. Most CEOs ask me this as their first question: "Should I be worried about losing control over where my data is stored?"

Although this issue receives a lot of attention from outsiders, the cloud's security ramifications are much more extensive than that.

- Legal actions that don't involve you cause security issues.
- Why virtualization was not taken into consideration when many of the regulations and standards that control your IT infrastructure were developed.
- In the cloud, the concept of perimeter security is completely meaningless.
- User credential management goes beyond simple identity management.
- Security in the cloud can truly be a challenge, just like it is in many other areas.

3.16 DATA PROTECTION

How you manage physical access to the servers that power your network is defined by physical security. There are still physical security limitations with the cloud. After all, somewhere there are active servers. You should be aware of the physical security standards used by the cloud provider you choose as well as the steps you need to take to protect your systems from physical threats [18].

3.17 DATA MANAGEMENT

The fact that your data is stored on someone else's computers separates the cloud significantly from traditional data centers. A portion of that gap may have been closed by businesses that outsourced their data centers to managed services providers, but cloud services add the impossibility of physically accessing the servers that house their data [19]. Although the significance of this transition is an emotionally charged issue, it does provide some serious commercial issues. The major issue is that your operations and your data could be compromised by circumstances that have nothing to do with your organization. For instance, any of the following circumstances could harm your infrastructure: The cloud provider files for bankruptcy, at which point its servers are seized [20].

3.18 LOCK EVERYTHING UP

Your data is kept someplace in the cloud, but you don't know where. But you are aware of several fundamental conditions:

- Your data is stored in a guest operating system of a virtual machine, and you have access control over that data.
- Other virtual hosts are not able to see the network traffic used to transfer data between instances.
- The majority of cloud storage services by default make data access private. But, a lot of them, including Amazon S3, let you make that data public.

3.19 YOUR NETWORK TRAFFIC USING ENCRYPTION

Regardless of how bad your existing security procedures are, network traffic is probably encrypted, at least in part. Being unable to sniff the traffic of other virtual servers is a wonderful feature of the Amazon cloud. Moreover, Amazon might release a future feature that makes this security measure unnecessary. So, not only web traffic but all network traffic should be encrypted [21].

3.20 AFFIRM BACKUP ENCRYPTION

You should use strong cryptography, such as PGP, to encrypt your data when you bundle it for backups. Then, you may safely keep information in an environment that is either fully unsafe or only partially secure, like Amazon S3.

Processor usage increases during encryption. Hence, before transferring the backups to your cloud storage system, I advise moving your plain-text files to a temporary backup server whose sole purpose is to conduct encryption [22]. The use of a backup server allows you to keep your cloud storage access credentials on a single higher-security machine rather than distributing those credentials, which not only prevents your application server and database server CPUs from being overworked [23].

3.21 MAKE YOUR FILE SYSTEMS SECURE

Each virtual server you administer will mount block storage devices or ephemeral storage devices (like the /mnt partition on Unix EC2 instances). It failed to encrypt the transient data in storage device. In an EC2 context, devices provide only a very low risk because the EC2 Xen system zeros out that storage when your instance shuts down. But, block storage device snapshots remain unencrypted in Amazon S3 unless you take additional steps to do so [24].

3.22 CONCLUSION

In conclusion, cloud computing and virtualization represent a paradigm shift in the way organizations conceive, deploy, and manage their IT infrastructures. The integration of cloud computing services, coupled with the inherent advantages of virtualization, offers businesses unparalleled opportunities for growth and efficiency. The scalability, cost-effectiveness, and accessibility provided by cloud computing empower organizations to adapt to dynamic market conditions seamlessly. Virtualization's ability to decouple software from underlying hardware enhances resource utilization, streamlines provisioning, and enables the creation of isolated environments, fostering a more agile and responsive IT ecosystem. As organizations increasingly migrate toward cloud-based solutions, the importance of virtualization in creating a foundation for these services becomes more evident. However, challenges such as security concerns, data privacy, and interoperability must be navigated to unlock the full potential of cloud computing and virtualization. Future advancements in these technologies promise even greater innovation, offering new possibilities for businesses to leverage as they continue to evolve in an ever-changing digital landscape. In essence, the journey toward cloud computing and virtualization is transformative, and organizations that strategically embrace and integrate these technologies are better positioned to thrive in the dynamic and competitive landscapes of the digital era. The synergy between cloud computing and virtualization represents not only a technological evolution but a strategic imperative for enterprises seeking sustainable growth and agility in an increasingly digital world.

REFERENCES

1. The NIST definition of cloud computing: recommendations of the National Institute of Standards and technology NIST Special Publication 800–145, P Mell, T Grance—Retrieved January 2011.
2. Lombardi, Flavio, and Roberto Di Pietro. "Secure virtualization for cloud computing." *Journal of Network and Computer Applications*, vol. 34, no. 4, 1113–1122, 2011.
3. Sato, Miyuki. "Creating next generation cloud computing based network services and the contributions of social cloud operation support system (oss) to society." In *2009 18th IEEE International Workshops on Enabling Technologies: Infrastructures for Collaborative Enterprises*, pp. 52–56. IEEE, 2009.
4. Wang, Lizhe, Jie Tao, Marcel Kunze, Alvaro Canales Castellanos, David Kramer, and Wolfgang Karl. "Scientific cloud computing: Early definition and experience." In *2008 10th IEEE International Conference on High Performance Computing and Communications*, pp. 825–830. IEEE, 2008.
5. Bouayad, Anas, Asmae Blilat, Nour elhouda Mejhed, and Mohammed El Ghazi. "Cloud computing: security challenges." IEEEComputer Society, 2012.

6. Tomar, Ashvin Singh, and Nagesh Salimath. "The help of cloud computing, the efficiency of virtual classes conducted in government colleges of Madhya Pradesh."

7. Kretzschmar, Michael, and S Hanigk. "Security management interoperability challenges for collaborative clouds." In *Systems and Virtualization Management (SVM), 2010, Proceedings of the 4th International DMTF Academic Alliance Workshop on Systems and Virtualization Management: Standards and the Cloud*, pp. 43–49, October 25–29, 2010. ISBN:978-1-4244-9181-0, DOI:10.1109/SVM.2010.5674744.

8. Amazon, A. W. S. "Amazon web services overview of security processes." 2015.

9. Grimes, Justin M., Paul T. Jaeger, and Jimmy Lin. "Weathering the storm: The policy implications of cloud computing." (2009).

10. Wu, Hanqian, Yi Ding, C. Winer, and Li Yao. "Network security for virtual machines in cloud computing,‖ 5th Int'l Conference on Computer Sciences and Convergence Information Technology." *Seoul*, November: 18–21.

11. Ma, Y., H. Wang, J. Dong, Y. Li, and S. Cheng. "Efficient live migration of virtual machine with memory exploration and encoding." In *Proc. CLUSTER*, September 2012, pp. 610–613.

12. Jin, H. et al. "MECOM: Live migration of virtual machines by adaptively compressing memory pages." *Future Generation Computer Systems*, vol. 38, pp. 23–35, Sep. 2014.

13. Kaur, Ramandeep, and Sumit Chopra. "Virtualization in cloud computing: a review." *International Journal of Scientific Research in Computer Science, Engineering and Information Technology (IJSRCSEIT)*, vol. 6, no. 4, pp. 01–05, July-August 2020. ISSN: 2456-3307.

14. Youry, K., and V. Volodymyr. "Cloud computing infrastructure prototype for university education and research." WCCCE '10, May 78, 2010, Kelowna, Canada.

15. *(PDF)* Virtualization in cloud computing: a review. Available from: https://www.researchgate.net/publication/346816485_Virtualization_In_Cloud_Computing_A_Review [accessed Apr 10, 2023].

16. Ouahabi, Sara, Ahmed Eddaoui, El Houssine Labriji, Elhabib Benlahmar, and Kamal El Guemmat. "Implementation of a novel educational modeling approach for cloud computing." *International Journal of Emerging Technologies in Learning (Online)*, vol. 9, no. 6, 49, 2014.

17. Justin, C., B. Ivan, K. Arvind, and A. Tom. "Seattle: a platform for educational cloud computing." SIGCSE09, March, 2009, Chattanooga, Tennessee, USA. 2009.

18. Bala, P. Shanthi. "Intensification of educational cloud computing and crisis of data security in public clouds." *International Journal on Computer Science and Engineering*, vol. 2, no. 3, 741–745, 2010.

19. CJB, R., and N. Evans. "A proposal for the adoption and use of cloud computing in secondary education in South Africa " In *11th DIS Annual Conference 2010, 2–3 September*, Richardsbay, University of Zululand, South Africa.

20. Armbrust, Michael, Armando Fox, Rean Griffith, Anthony D. Joseph, Randy H. Katz, Andrew Konwinski, Gunho Lee, et al. *Above the clouds: A Berkeley view of cloud computing*, vol. 17. Technical Report UCB/EECS-2009-28, EECS Department, University of California, Berkeley, 2009.

21. Al Noor, S., G. Mustafa, S. Chowdhury, Z. Hossain, and F. Jaigirdar. "A proposed architecture of cloud computing for education system in Bangladesh and the impact on current education system." *International Journal of Computer Science and Network Security (IJCSNS)*, vol. 10, no. 10. 2010.

22. Muniasamy, Vasanthi, Intisar Magboul Ejalani, and M. Anandhavalli. "Moving towards virtual learning clouds from traditional learning: Higher educational systems in India." *International Journal of Emerging Technologies in Learning (Online)*, vol. 9, no. 9, 70, 2014.

23. Tomar, Ashvin Singh, and Nagesh Salimath. "THE help of cloud computing, the efficiency of virtual classes conducted in government colleges of Madhya Pradesh."

24. Muniasamy, Vasanthi, Intisar Magboul Ejalani, and M. Anandhavalli. "Moving towards virtual learning clouds from traditional learning: Higher educational systems in India." *International Journal of Emerging Technologies in Learning (Online)*, vol. 9, no. 9, 70, 2014.

Chapter 4

Deep reinforcement learning based framework for tactical drone deployment in rigorous terrains

From modeling to real-world implementation

Shekhar Tyagi[a] and Abhishek Tyagi[b]

[a]Department of Computer Science and Engineering, IIMT College of
Engineering, Greater Noida, Uttar Pradesh, India
[b]Department of Computer Science and Engineering, KIET Group of
Institutions, Ghaziabad, Uttar Pradesh, India

4.1 INTRODUCTION

Deep reinforcement learning (DRL) represents the intersection of reinforcement learning with deep learning [1,2] (Figure 4.1).

4.1.1 Basics of reinforcement learning

At its essence, reinforcement learning revolves around an agent's ability to engage with its surroundings in order to optimize a cumulative reward [1].

4.1.1.1 Key components

- **Agent:** This is the decision-maker in the RL framework.
- **Environment:** The setting in which the agent operates.

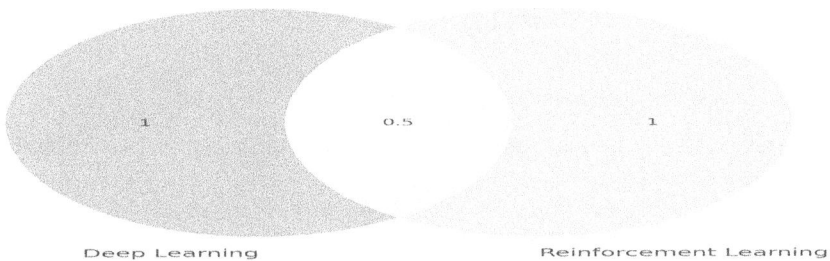

Deep Reinforcement Learning Intersection

Deep Learning   Reinforcement Learning

Figure 4.1 Venn diagram illustrating the overlap between deep learning and reinforcement learning, leading to deep reinforcement learning.

- State *s*: The state is a representation of the environment at a given time.
- Action *a*: Actions refer to the selections or determinations an agent makes.
- Reward *r*: Upon executing an action, the agent is granted a reward by the environment.

4.1.2 The RL problem: Policy, value function, Q-function [3]

4.1.2.1 Policy π

A policy defines a strategy or a mapping that defines how an agent should act in a given state [4].

4.1.2.2 Valuemetric function (Vm-function)

The valuemetric function, denoted as $V^{\pi}(s)$, represents the expected return an agent can achieve beginning from state *s* and following policy π [5].

4.1.2.3 Action-valuemetric function (av-Q-function)

The av-Q-function, denoted as $Q^{\pi}(s, a)$, represents the expected return of taking action
a in state *s* and then following policy π [6].

4.2 DEEP LEARNING—PRIMERS

Deep learning, a branch of machine learning, employs multi-layered neural networks (deep structures) to examine diverse aspects of data [2].

4.2.1 Neural networks

A neural network comprises a sequence of algorithms that discern inherent connections within a data set [2].

4.2.2 Back-propagation

Back-propagation is a supervised learning technique employed to train neural networks [2].

4.2.3 Activation functions

Activation functions infuse non-linearity into the network [2].

4.2.4 Convolutional neural nets (CNNs)

CNNs belong to a category of deep neural networks particularly adept at activities such as image identification and categorization [2].

4.2.5 Recurrent neural nets (RNNs)

RNNs are tailored for sequential information and operations that need recollection of preceding inputs [2].

4.2.6 Deep Q-learning overview

Q-learning is an off-policy, model-independent reinforcement learning technique [3].

The Q-value $Q(s, a)$ represents the expected future rewards an agent can achieve from state s by taking action a and following an optimal policy thereafter.

$$Q(s, a) \leftarrow Q(s, a) + \alpha^h r + \beta \max_{a'} Q(s', a') - Q(s, a)^i \qquad (4.1)$$

Where:

- α is the factor for learning rate.
- β is the rate of discount factor.
- r is an immediate reward measure after initiating action a over state s.
- s' is next state.
- $\max_{a'} Q(s', a')$ is the estimating value of future rewards.

4.2.6.1 Deep Q-networks (DQNs)

The main challenge with Q-learning is that it doesn't scale well to environments with a large volume of states. DQNs use neural networks to approximate Q-values given a state.

4.2.6.2 Experienced replaying and targeting networks

1. Experienced Replay: Instead of learning from some recent experience, the agent stores its experience and later samples a batch from this buffer to train the network.
2. Target Networks: DQN uses a separate network to estimate the future Q-values in the update equation.

4.3 POLICY GRADIENT METHODS

4.3.1 Motivation for policy-based methods

Reinforcement learning algorithms can generally be divided into two main types: value-based and policy-based approaches. The primary motivations for policy-based methods are:

- Stochastic Policies
- Continuous Action Spaces
- Simplicity

4.3.2 REINFORCE algorithm

The REINFORCE algorithm is the foundational policy gradient method. The algorithm consists of the following steps:

1. Objective to learn a policy π_θ that maximizes the expected reward:

$$J(\theta) = E\pi_\theta[R] \tag{4.2}$$

2. Leverage the policy gradient theorem:

$$\nabla_\theta J(\theta) = E_{\pi_\theta}[\nabla_\theta \log \pi_\theta(a|s) \times R] \tag{4.3}$$

3. Update the policy parameters.

4.3.3 Actor-critic methods

Actor-critic methods combine the strengths of both value-based and policy-based methods. The primary benefit is that the critic can provide a less noisy and more stable estimate of the expected future rewards.

4.4 ADVANCED TECHNIQUES IN DRL

4.4.1 Double DQN

Double DQN introduces a simple modification to address the issue of overestimation in DQNs. The update rule is:

$$\arg\max Q(s, a) \leftarrow r + \gamma Q'(s', \arg\max a Q(s', a)) \tag{4.4}$$

4.4.2 Dueling DQN

Dueling DQN introduces a new architecture for the Q-network. The Q-value is then computed as:

$$\text{mean}Q(s, a) = V(s) + (A(s, a) - \text{mean}A(s, a)) \tag{4.5}$$

4.4.3 Proximal policy optimization (PPO)

PPO is designed to improve data efficiency and stability in training. The objective function in PPO is:

$$L(\vartheta) = \min \frac{\pi_\theta(a|s)}{\pi_{\theta_{\text{old}}}(a|s)} A_{\pi_{\theta_{\text{old}}}}(s, a), \text{clip} \frac{\pi_\theta(a|s)}{\pi_{\theta_{\text{old}}}(a|s)}, 1 - \epsilon, 1 + \epsilon A_{\pi_{\theta_{\text{old}}}}(s, a) \tag{4.6}$$

4.4.4 Hindsight experience replay (HER)

HER is a technique that makes use of failed trajectories to learn more efficiently.

4.5 CHALLENGES AND OPEN PROBLEMS IN DRL

4.5.1 Exploration vs. exploitation

One of the central dilemmas in reinforcement learning is the trade-off between exploration and exploitation. Balancing these is challenging and has implications for the effectiveness of the learning process.

4.5.2 Sample inefficiency

DRL often requires a significant number of samples to learn an effective policy. This inefficiency is particularly problematic in real-world scenarios.

4.5.3 Non-stationarity and instability

Learning in DRL is non-stationary, introducing complexities that can destabilize learning and make training more challenging.

4.5.4 Transfer learning and meta-Learning in DRL

Transfer learning and meta-learning are emerging as critical tools in DRL for improving data efficiency and adaptability.

4.5.5 Multi-agent reinforcement learning (MARL)

MARL focuses on scenarios where multiple agents learn simultaneously. This presents challenges like coordination, non-stationarity, and issues of communication and cooperation versus competition.

4.5.6 Ethics and safety in DRL applications

As DRL finds increasing applications in real-world scenarios, concerns about its ethical use and safety are rising. Questions of fairness, transparency, and accountability need to be addressed, especially in critical sectors like healthcare and finance.

4.6 DRL FOR TACTICAL DRONE DEPLOYMENT

4.6.1 Introduction and background

The defense forces of various nations have been exploring advanced technological solutions to enhance their surveillance capabilities. DRL offers promising solutions for autonomous systems, including drones.

4.6.2 State (environment)

The state in reinforcement learning provides a comprehensive snapshot of the current situation for the agent. In the context of drone deployment, this encompasses several factors.

4.6.2.1 Terrain data

Every location the drone might fly over will have associated terrain data (Figures 4.2–4.4).

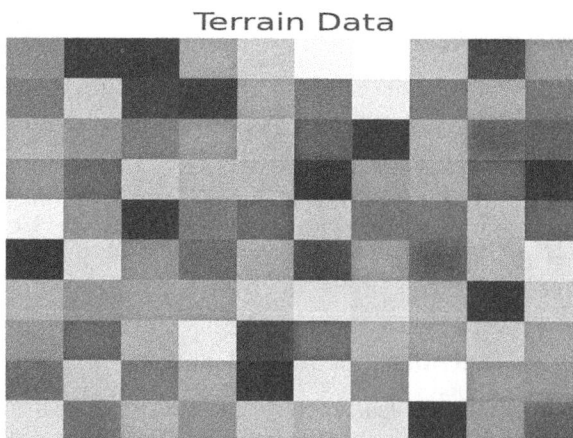

Figure 4.2 Terrain heatmap.

4.6.3 The deployment framework

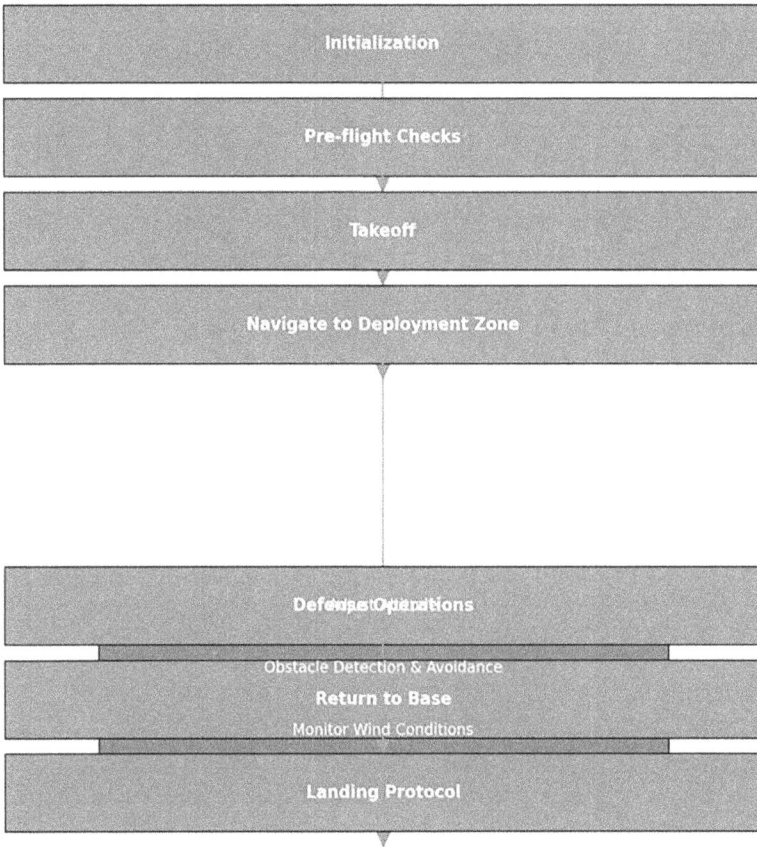

Figure 4.3 Block diagram representation of defense drone deployment process.

4.6.4 Key highlights

- Initialization
- Pre-flight Checks
- Takeoff
- Navigate to Deployment Zone
- Defense Operations
- Return to Base
- Landing Protocol

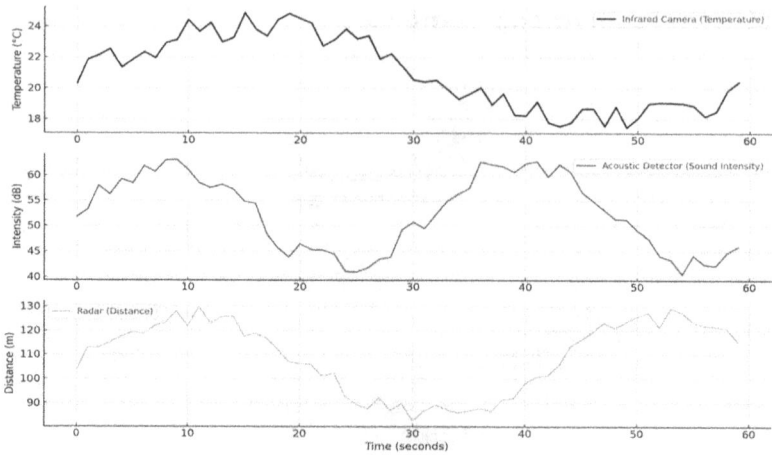

Figure 4.4 Drone sensor readings.

4.6.5 Sensor readings

The sensor is **reading that intensity and translating it into a reading.** This ability to trace an external light source is interesting, but it's probably not the most amazing thing you can do with this sensor. There is another feature of this device: Not only does it detect light, but it emits some light as well.

4.6.6 Procedure

- Data Generation
- Time Series
- Infrared Camera
- Acoustic Detector
- Radar
- Visualization

Algorithm 1 DeployDefenseDrone

1: **function** DEPLOYDEFENSEDRONE(target coordinates)
2:    INITIALIZE   drone
3:    SET safe altitude = GETSAFEALTITUDEFROMTOPOGRAPHICALMAP
4:    ESTABLISH COMMUNICATION with control station
5:    **if** NOT PREFLIGHTCHECKS **then**
6:       **return** "Pre-flight checks failed!"
7:    **end if**
8:    ASCEND TO safe altitude
9:    CALIBRATE SENSORS()
10: **end function**

Algorithm 2 Navigate Deployment Zone

1: **while** NOT AtTarget(target coordinates) **do**
2:     current altitude = GETCURRENTALTITUDE
3:     **if** current altitude ¡ safe altitude **then**
4:         ADJUST ALTITUDE TO safe altitude
5:     **end if**
6:     NAVIGATE TOWARDS target coordinate
7: **end while**

Algorithm 3 Obstacle Detection and Avoidance

 1:   **if** OBSTACLEDETECTED **then**
 2:     AVOID OBSTACLE()
 3: **end if**
 4: **function** OBSTACLEDETECTED
 5:     Uses radar and other sensors to detect obstacles
 6: **end function**
 7: **function** AVOIDOBSTACLE
 8:     Adjust flight path to avoid detected obstacles
 9: **end function**
10: CHECK obstacle detection system
11: **if** ANY check failed **then**
12:     **return** False
13: **else**
14:     **return** True
15: **end if**
16: wind conditions = CHECK WIND CONDITIONS()
17: **if** wind conditions ARE DANGEROUS **then**
18:     HOVER OR FIND SAFE LANDING()
19: **end if**

Algorithm 4 Defense Operations and Landing Protocols

 1: **while** ON MISSION **do**
 2:     SCAN FOR THREATS()
 3:     **if** THREATDETECTED **then**
 4:         ALERT CONTROL STATION()
 5:         RECORD FOOTAGE()
 6:         MARK GPS LOCATION()
 7:     **end if**
 8:     **if** BATTERYLOW **then**
 9:         **break**
10:     **end if**
11: **end while**

12: RETURN TO BASE()
13: **if** AtBase() **then**
14:    LAND SAFELY()
15:    TRANSMIT MISSION DATA()
16: **end if**

Algorithm 5 PreFlightChecks

1: **function** PREFLIGHTCHECKS
2:    CHECK battery level
3:    CHECK sensors operational
4: **end function**

4.7 PSEUDO CODES OF THE ENTIRE FRAMEWORK

4.7.1 *Pseudo code for the process INIT DEPLOYMENT*
4.7.2 *Pseudo code for the process NAVIGATE DEPLOYMENT ZONE*
4.7.3 *Pseudo code for the process*
        OBSTACLE DETECTION AVOIDANCE()
4.7.4 *Pseudo code for the process*
        DEFENCE OPERATIONS LANDING PROTOCOLS
4.7.5 *PreFlightChecks Function*

4.8 DRONE STATUS

This includes metrics such as the drone's battery level, communication signal strength, current speed, and direction (Figure 4.5).

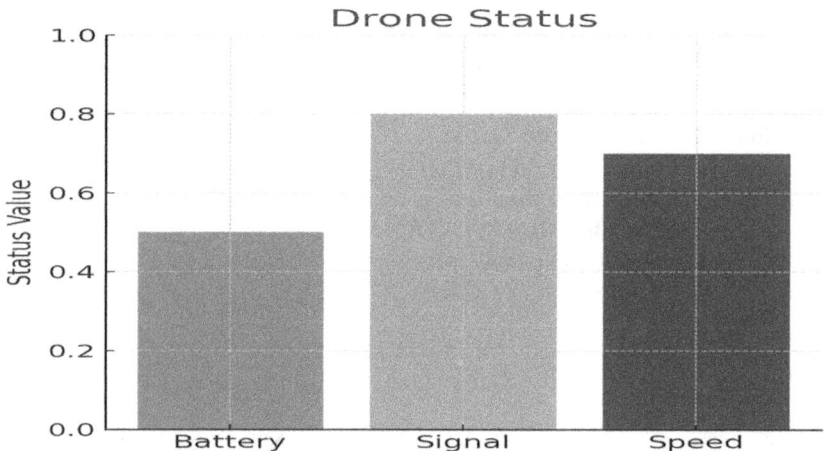

Figure 4.5 Drone metrics status.

Threat Data (Heatmap)

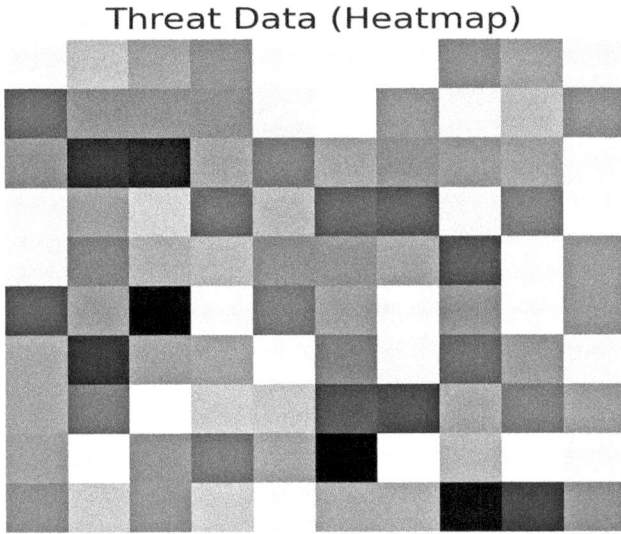

Figure 4.6 Threat data heatmap.

4.9 THREAT DATA

Historical data about where threats have been encountered can guide the drone's patrolling patterns. Real-time threat data can influence immediate decisions (Figure 4.6).

4.10 ACTION

In the DRL framework, actions dictate the drone's movements and reactions to its environment. Actions for our drone can be grouped into several categories (Figure 4.7):

- Movement: The drone can change its altitude, direction, or speed based on its current state and objectives.
- Sending alerts: Whenever the drone detects something that matches the threat profile, it sends an alert to the command center.
- Return to base: Especially in situations of low battery or malfunction.

Send Alert and Return to Base

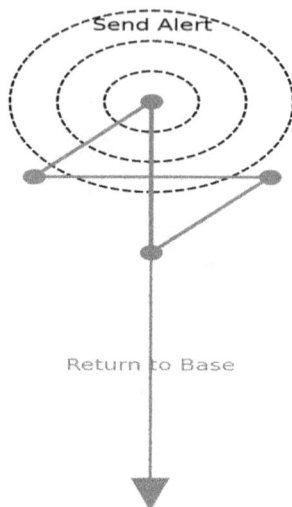

Figure 4.7 Send alert and return to base signaling.

4.11 REWARD FUNCTION

The reward function plays a pivotal role in reinforcement learning, providing feedback based on the outcomes of the drone's actions (Figure 4.8).

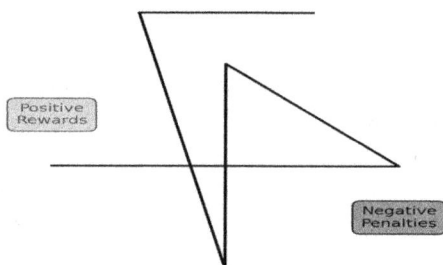

Figure 4.8 Reward function visualization.

4.12 DRL AGENT

The DRL agent is the "brain" of the drone, making decisions based on the input state. The agent typically comprises a neural network that approximates the Q-function.

4.13 POLICY

The policy defines how the drone chooses its actions. Initially, it might use an exploratory policy and later act based on its learned strategy (Figure 4.9).

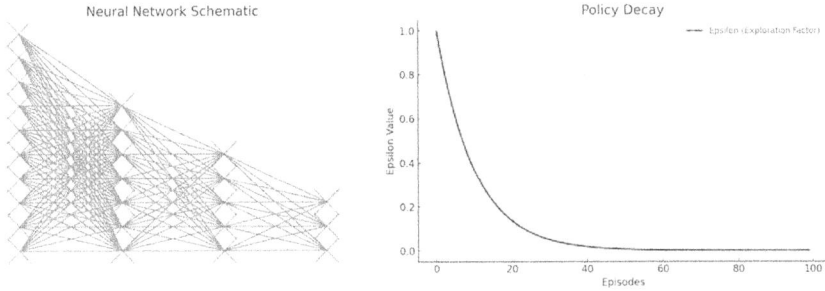

Figure 4.9 Schematic neural network and policy decay.

4.14 TRAINING

Before real-world deployment, the DRL agent is trained in a simulated environment, often referred to as a "digital twin", that mimics the actual environment it will operate in (Figure 4.10).

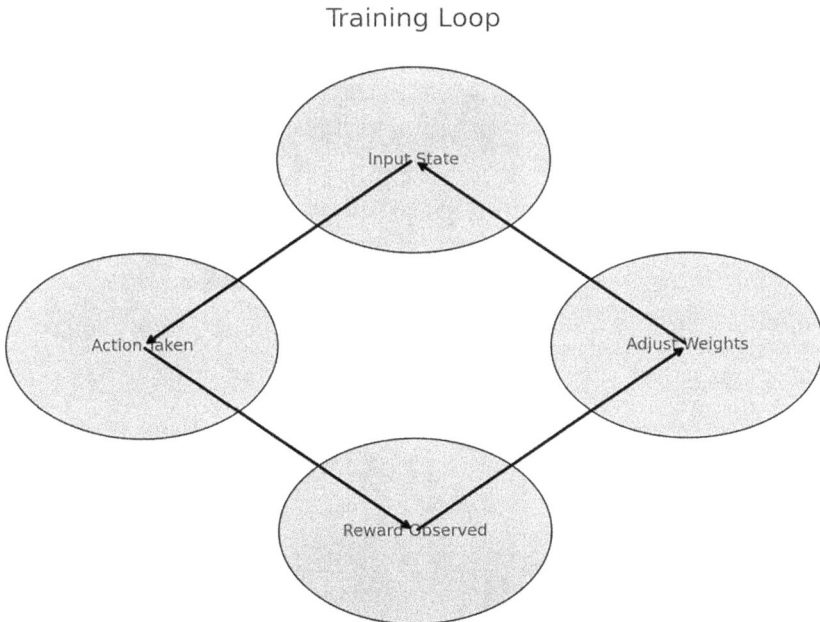

Figure 4.10 Training loop visualization.

4.15 DEPLOYMENT

After rigorous training and validation, the DRL agent is embedded into the real drones. These drones then operate in the actual environment, leveraging the knowledge gained during training (Figures 4.11 and 4.12).

Deployment: Drone in Operation

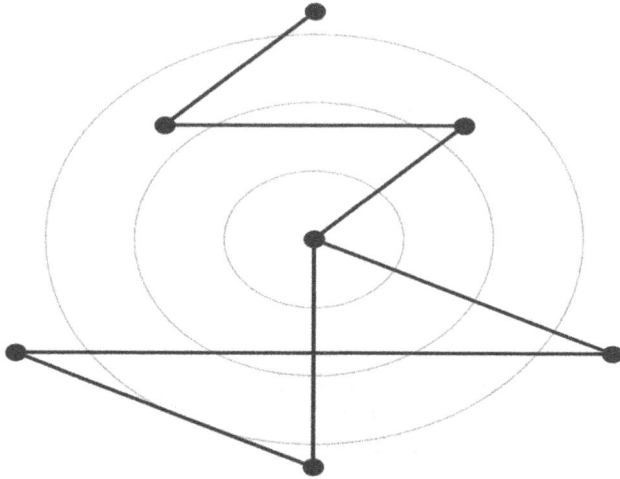

Figure 4.11 Drone deployment visualization.

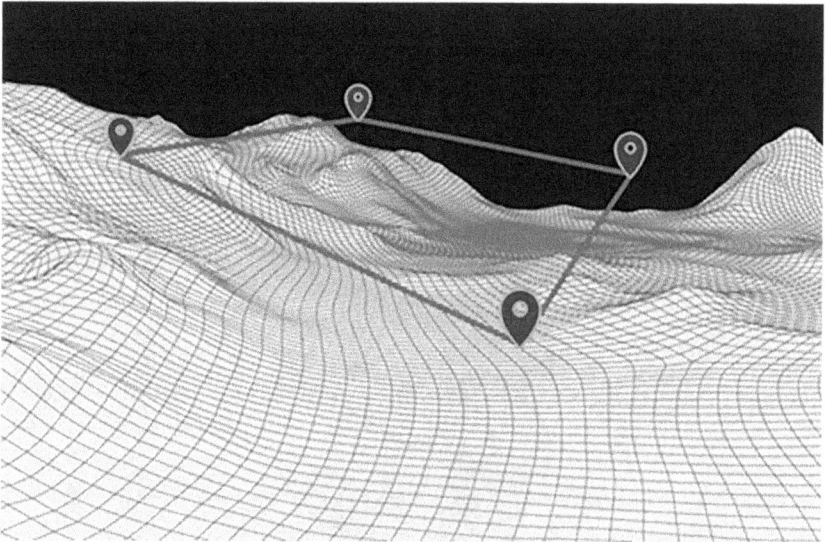

Figure 4.12 Lidar mapping of real site deployment based on latitude and longitude data of a terrain.

4.16 CONCLUSION

DRL holds significant potential for enhancing defense operations, especially in the realm of autonomous drone deployment. By integrating advanced AI techniques with real-world systems, defense forces can achieve improved surveillance coverage, faster threat response times, and overall better operational efficiency.

DISCLOSURE STATEMENT

We declare that we do not have any conflicts of interest.

REFERENCES

1. Sutton, R. S., & Barto, A. G. (2018). *Reinforcement learning: An introduction*. MIT Press.
2. Goodfellow, I., Bengio, Y., & Courville, A. (2016). *Deep learning*. MIT Press.
3. Mnih, V., Kavukcuoglu, K., Silver, D., Graves, A., Antonoglou, I., Wierstra, D., & Riedmiller, M. (2013). Playing Atari with deep reinforcement learning. *arXiv preprint arXiv:1312.5602*.
4. Williams, R. J. (1992). Simple statistical gradient-following algorithms for connectionist reinforcement learning. *Machine Learning*, 8(3-4), 229–256.
5. Schulman, J., Wolski, F., Dhariwal, P., Radford, A., & Klimov, O. (2017). Proximal policy optimization algorithms. *arXiv preprint arXiv:1707.06347*.
6. Henderson, P., Islam, R., Bachman, P., Pineau, J., Precup, D., & Meger, D. (2018). Deep reinforcement learning that matters. In *Proceedings of the AAAI Conference on Artificial Intelligence*, 32(1).

Chapter 5

Elevating healthcare services

The integration of cloud with fog computing

Tarun Kumar Vashishth, Vikas Sharma,
Kewal Krishan Sharma, and Bhupendra Kumar
School of Computer Science and Applications, IIMT University, Meerut,
Uttar Pradesh, India

5.1 INTRODUCTION

In today's rapidly evolving healthcare landscape, the pursuit of improved patient care, enhanced clinical outcomes, and efficient healthcare service delivery is paramount. The healthcare segment is undergoing a profound transformation, with upgraded technologies that promise to revolutionize the way healthcare services are provided, managed, and accessed. Among these transformative technologies, the integration of cloud computing and fog computing stands out as a promising paradigm shift with the potential to elevate healthcare services to unprecedented levels of quality, accessibility, and efficiency. The healthcare industry faces an array of challenges, including the ever-increasing volume of patient data, the need for real-time monitoring and decision-making, the demand of personalized treatment, and the imperative to reduce costs while maintaining or even enhancing patient care. Traditional healthcare systems, primarily based on localized, on-premises infrastructure, often struggle to address these challenges comprehensively. In this context, cloud and fog computing emerge as powerful tools capable of reshaping the healthcare landscape. Cloud computing and fog computing in healthcare services is shown in Figure 5.1.

CC has already made significant inroads into healthcare by providing scalable storage, computational resources, and data analytics capabilities. It enables healthcare organizations to store vast amounts of patient data securely, harness the power of AI and ML for diagnosis and treatment recommendations, and facilitate collaborative research on a global scale. However, cloud computing also presents certain limitations, including latency issues, data privacy concerns, and the dependency on stable internet connectivity. Fog computing, a relatively recent paradigm in the realm of edge computing, complements cloud computing by extending its capabilities to the edge network, closer to where data is generated and consumed. This architecture alleviates the latency challenges associated with cloud computing and ensures real-time processing of critical healthcare data. Fog computing leverages a distributed network of edge devices, such as expert medical

DOI: 10.1201/9781003461418-5

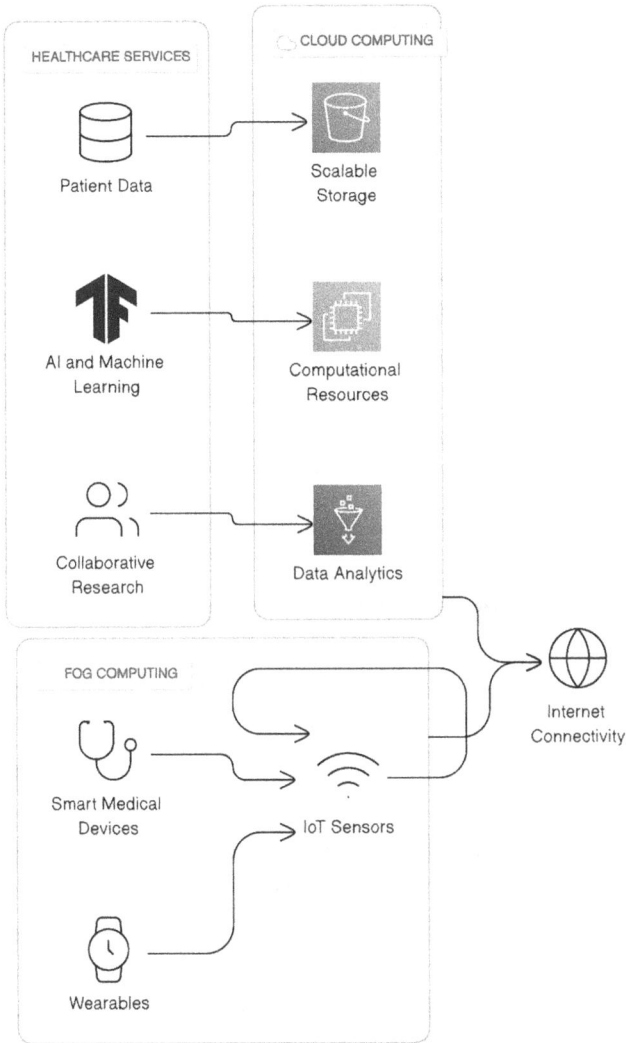

Figure 5.1 Cloud computing and fog computing in healthcare services.

devices, wearables, and IoT sensors, to give timely and context-aware services, making it particularly well suited for healthcare applications. The integration of fog and cloud computing in healthcare services represents a novel approach that holds immense potential. By combining the strengths of both paradigms, healthcare providers can harness the computational power and scalability of the cloud with an advantage from the lower latency, real-time analyzing capacity of fog computing. This integration not only enhances the quality of patient care but also optimizes resource utilization, reduces costs, and fosters innovation in healthcare delivery.

This research study aims to explore combination of cloud computing and fog computing in healthcare services comprehensively. It will delve into the technical aspects of this integration, including architecture, data management, security, and interoperability. Additionally, it will investigate the practical applications and use cases across various healthcare domains, such as telemedicine, remote patient monitoring, predictive analytics, and personalized medicine. Furthermore, this paper will address the ethical and regulatory considerations associated with the adoption of cloud and fog computing in healthcare, ensuring that patient privacy and data security are upheld. As the healthcare industry navigates the complexities of the digital age, the combination of cloud computing and fog computing emerges as a pivotal strategy to overcome existing challenges and pave the way for a more efficient, patient-centric, and data-based healthcare ecosystem. This chapter seeks to provide valuable insights and guidance to healthcare stakeholders, policymakers, and researchers alike, as they endeavor to leverage these transformative technologies for the betterment of healthcare services and, ultimately, the well-being of patients worldwide.

5.1.1 The evolving healthcare landscape

The healthcare segment is going through a remarkable change, guided by technological advancements and the expectations of patients and providers alike. Over the past decade, this industry has witnessed a profound paradigm shift, characterized by the acceptance of electronic health records (EHRs), the proliferation of wearable devices, and the growing reliance on telemedicine. This transformation has generated an unprecedented volume of data that is reshaping how healthcare is delivered, managed, and experienced.

Electronic Health Records (EHRs): One of the key drivers of change in the healthcare sector has been the widespread adoption of electronic health records (EHRs). EHRs replace traditional paper-based patient records with digital systems that store comprehensive patient data, and comprise of medical background, diagnoses, medications, and treatment plans. These transitions have not only improved the correctness and accessibility of patient data but have also enhanced the coordination of care among healthcare providers. EHRs facilitate seamless information sharing among different healthcare facilities, ensuring that medical professionals have access to up-to-date patient records, regardless of location. This interoperability has streamlined care delivery, reduced duplication of tests and procedures, and ultimately improved patient outcomes. See Figure 5.2, which shows electronic health records.

Proliferation of Wearable Devices: Another significant development in the evolving healthcare landscape is the proliferation of wearable devices. These devices, ranging from smart watches and fitness trackers to specialized medical wearables, continuously collect a wealth of health-related data. They monitor vital signs, track physical activity, and even detect

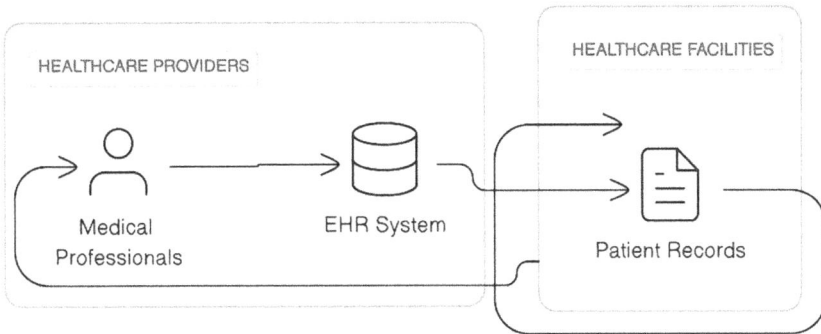

Figure 5.2 Electronic health records (EHRs).

irregularities in heart rhythms, providing individuals with valuable insights into their health and well-being. For healthcare providers, wearable devices offer an opportunity to gather real-time data that was previously unavailable. This data can be used for remote patient monitoring, early disease detection, and personalized treatment plans. Wearables devices enhanced the power of individuals to take an active role to maintain their health and enable healthcare professionals to intervene proactively, when necessary, potentially preventing complications and hospitalizations.

Growing Reliance on Telemedicine: Telemedicine has emerged as a transformative force in healthcare, particularly in response to the COVID-19 pandemic. This technology permits an ill person to consult with healthcare vendors remotely, removing the need for in-person visits in many cases. Telemedicine services have expanded access to care, reduced wait times, and provided a convenient and safe way for patients to receive medical advice and treatment. Furthermore, telemedicine has opened the door to telehealth monitoring, where patients can use connected devices to communicate important signs and health information to healthcare system on a real-time basis. This enables continuous monitoring of chronic conditions and post-operative care, improving patient outcomes and reducing the burden on healthcare facilities.

Harnessing Data for Better Patient Care: The exponential growth of data in the healthcare sector presents both opportunities and challenges. To unlock the full potential of this data, healthcare organizations require sophisticated solutions for data management, analytics, and decisions. Augmented analytics and ML algorithms can sift through vast data sets to identify trends, predict disease outbreaks, and optimize treatment plans. These tools enable healthcare providers to make evidence-based decisions, tailor interventions to individual patients, and allocate resources more efficiently.

However, this data-driven transformation also points to critical considerations regarding data security and ethics. Healthcare organizations must implement robust cybersecurity measures to protect sensitive patient information. Additionally, they must navigate the complex regulatory landscape to ensure compliance with privacy of data-related laws, such as the Health Insurance Portability and Accountability Act (HIPAA) in the USA.

5.2 CLOUD COMPUTING IN HEALTHCARE

Cloud computing has emerged as a transformative force in the healthcare sector, playing a pivotal role in elevating healthcare services through enhanced data storage, processing capabilities, and collaborative healthcare solutions. In the context of healthcare, cloud computing offers a scalable and flexible infrastructure that allows healthcare providers to efficiently manage and analyze vast amounts of patient data. The ability to store and access data in the cloud facilitates seamless collaboration among healthcare professionals, enabling them to make more informed decisions and provide better patient care. Electronic health records (EHRs), medical imaging, and other data-intensive applications benefit from the centralized storage and processing power of the cloud, streamlining workflows and improving overall healthcare efficiency.

However, despite its advantages, traditional cloud computing models face challenges in meeting the stringent requirements of healthcare applications, particularly when it comes to low-latency processing and real-time data analysis. This is where the integration of cloud computing with fog computing becomes crucial. Fog computing extends the cloud model by bringing computation, storage, and communication capabilities closer to the edge of the network. In healthcare scenarios, this means that data processing can occur at or near the source of the data, reducing latency and improving response times. This is particularly significant in applications like remote patient monitoring, where timely data analysis is paramount for making critical decisions about patient care. The integration of cloud computing with fog computing in healthcare enhances resource management and addresses the limitations of relying solely on a centralized cloud infrastructure. Fog computing facilitates the processing of data from a multitude of Internet of Things (IoT) devices, wearables, and medical sensors at the network edge. This distributed approach optimizes resource utilization by handling data locally, reducing the need for continuous high-bandwidth connections to the cloud. In healthcare, where massive amounts of data are generated continuously, this distributed computing model improves overall system efficiency. Additionally, fog computing enhances data security and privacy by minimizing the transmission of sensitive information to the cloud, ensuring that critical patient data remains closer to its source.

Furthermore, the hybrid cloud-fog model in healthcare resource management involves strategic allocation of tasks between the cloud and fog layers. Cloud resources are leveraged for computationally intensive tasks, such as running complex analytics and machine learning algorithms on large data sets. Fog computing, on the other hand, takes care of real-time data processing and edge analytics. This collaboration between cloud and fog computing allows healthcare organizations to achieve a balance between high-performance computing capabilities and the responsiveness needed for time-sensitive applications.

In conclusion, the integration of cloud computing with fog computing represents a paradigm shift in healthcare services, addressing the evolving needs of the industry. This hybrid model not only overcomes the limitations of traditional cloud computing but also enhances resource management, data security, and overall system efficiency. As healthcare continues to embrace digital transformation, the integration of cloud with fog computing stands out as a key enabler, fostering innovation and elevating the quality of patient care. The collaborative power of these technologies ensures that healthcare professionals have access to the right information at the right time, ultimately leading to improved patient outcomes.

5.2.1 Data storage and accessibility

Cloud computing has fundamentally transformed the way healthcare organizations handle data. One of its most significant contributions is the revolution in data storage and accessibility. EHRs, diagnostic images, patient histories, and other critical healthcare data are now securely hosted in the cloud. This shift has far-reaching implications for the healthcare industry.

Accessibility: The cloud's accessibility means that authorized personnel, including doctors, nurses, and administrative staff, can retrieve vital patient information from virtually anywhere with an internet connection. This seamless access to EHRs and other clinical data has immense advantages. For instance, in emergency situations, healthcare vendors can access a patient's medical past details and previous treatments, enabling them to make more informed decisions promptly. This real-time access enhances care coordination among different providers, reduces the risk of errors, and ultimately improves patient outcomes.

Cost-Efficiency: Cloud-based storage eliminates the need for on-premises data centers, reducing the associated infrastructure and maintenance costs. Healthcare organizations can allocate resources more efficiently, directing funding toward patient care and innovation rather than IT infrastructure. This cost-effectiveness is particularly beneficial for smaller healthcare facilities and those with limited budgets.

5.2.2 Scalability

Healthcare is a dynamic field with fluctuating data processing needs. Patient admissions, clinical trials, research projects, and other factors can significantly impact the volume and complexity of data that healthcare organizations must handle. Cloud computing offers unparalleled scalability, which allows healthcare providers to adapt their computational resources to meet changing demands.

Resource Flexibility: Cloud computing services enable healthcare organizations to extend their IT resources up or down in respect to requirement. For instance, during a flu outbreak, a hospital can quickly increase its server capacity to handle a surge in patient data. Conversely, during quieter periods, they can reduce resources to avoid unnecessary operational costs. This flexibility optimizes resource utilization and ensures that computing power aligns with specific needs, maximizing efficiency. See Figure 5.3, which shows cloud computing with EHRs.

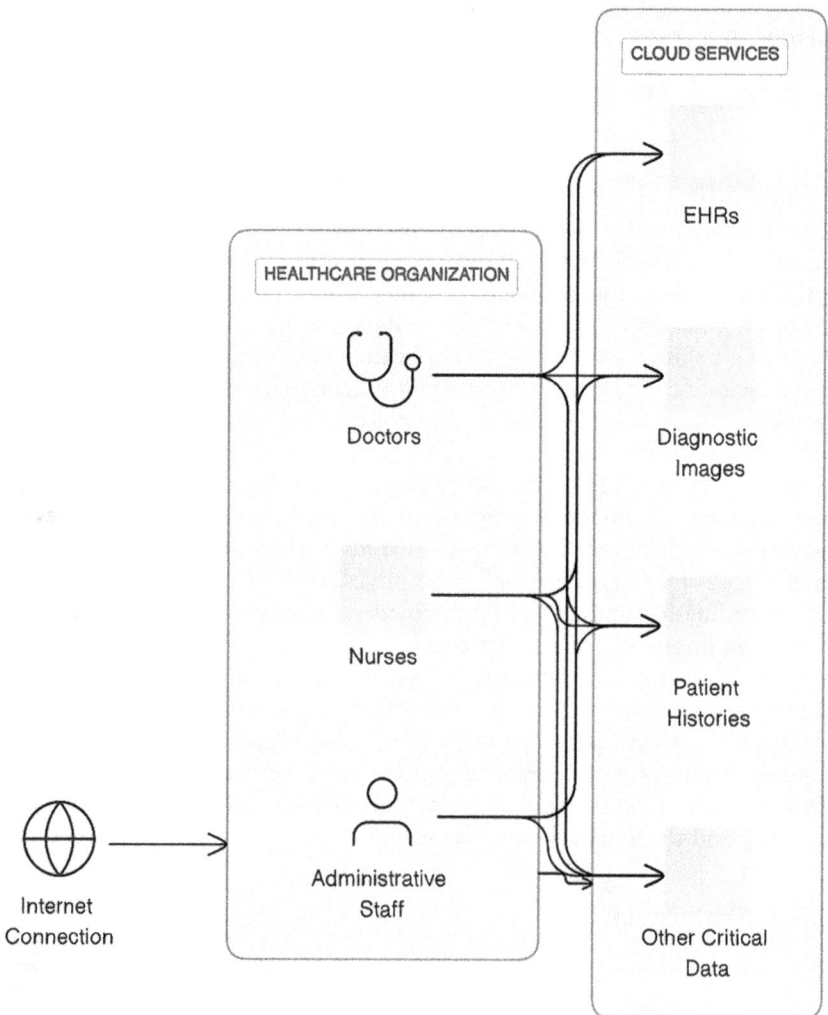

Figure 5.3 Cloud computing with EHRs.

5.2.3 Data analytics with machine learning

Cloud serves as an ideal platform for data analytics and machine learning applications in healthcare. Leveraging cloud-based analytics tools and machine learning algorithms has become a game-changer for the industry, ushering in a new era of data-driven decision-making and patient care.

Actionable Insights: By utilizing cloud-based data analytics, healthcare providers can extract actionable insights from vast amounts of patient data. These insights encompass a wide range of applications, from predicting disease outbreaks to identifying high-risk patient populations. For instance, predictive analytics can help hospitals anticipate admissions of patient and allocate resources accordingly, strengthening resource management and patient care quality.

Disease Modeling: Cloud-based machine learning models have the capability to create sophisticated disease models. These models can predict disease progression, allowing for early intervention and personalized treatment plans. For instance, in cancer care, machine learning can analyze genetic data to determine the most effective treatment strategies for individual patients, minimizing side effects and optimizing outcomes.

Personalized Treatment: Cloud-based machine learning can process patient data, including medical past history, genetics, and treatment responses, to tailor treatment plans to individual patients. This personalized approach improves treatment effectiveness and minimizes adverse effects, ultimately enhancing patient satisfaction and outcomes. See Figure 5.4, which shows data analytics and machine learning applications in healthcare.

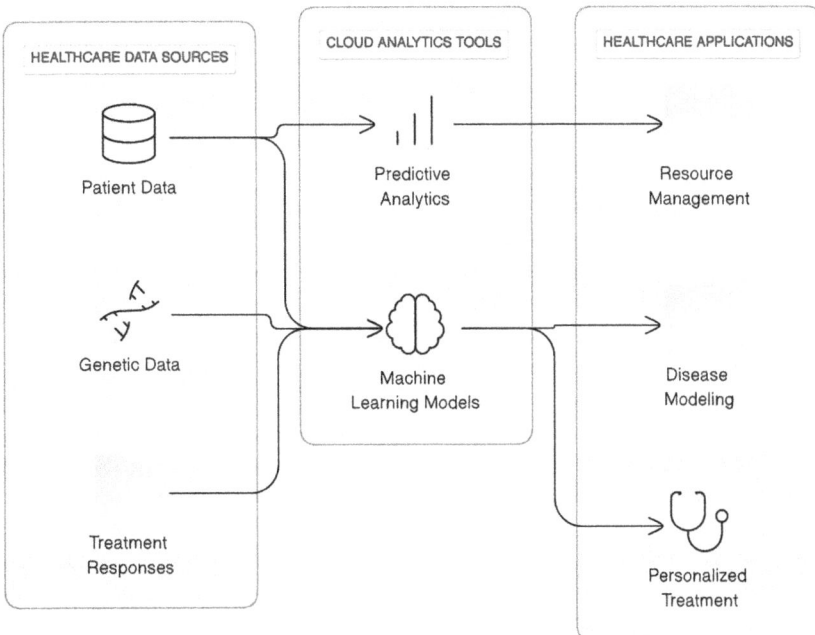

Figure 5.4 Data analytics and machine learning applications in healthcare.

5.3 FOG COMPUTING IN HEALTHCARE

Fog computing is proving to be a groundbreaking advancement in the healthcare sector, playing a pivotal role in elevating healthcare services by addressing the unique challenges posed by the industry's complex and dynamic nature. Fog computing extends the capabilities of cloud computing by bringing processing, storage, and communication closer to the edge of the network, directly where data is generated. In the context of healthcare, this distributed computing model offers significant advantages. One of the key benefits is the reduction of latency, a critical factor in healthcare applications where real-time data processing is essential. Unlike traditional cloud computing, fog computing enables healthcare organizations to process data locally at the network edge, ensuring rapid response times and facilitating time-sensitive tasks such as remote patient monitoring, emergency response systems, and critical decision-making.

The integration of fog computing with cloud computing in healthcare is a strategic move that optimizes resource management and enhances overall system efficiency. Fog computing allows for the processing of data from a multitude of Internet of Things (IoT) devices, wearables, and medical sensors at the edge, minimizing the strain on centralized cloud servers. This not only conserves bandwidth but also reduces the load on the cloud infrastructure, leading to more efficient resource allocation. Moreover, the proximity of fog computing to the data source enhances data security and privacy by limiting the transmission of sensitive information to the cloud, assuring patients that their critical health data remains closer to its origin.

In healthcare, where data volume and diversity are ever-expanding, fog computing contributes significantly to real-time data analytics. This is especially valuable in scenarios such as monitoring patient vitals, where instant analysis of data trends can prompt immediate intervention. Additionally, the integration of fog computing with cloud resources allows for a dynamic division of tasks. Heavy computational tasks that require substantial processing power, such as complex analytics and machine learning algorithms applied to extensive data sets, can be offloaded to the cloud. Meanwhile, fog computing handles the processing of data that demands low-latency responses, ensuring a balanced and efficient utilization of computing resources.

The collaboration between cloud and fog computing in healthcare is transformative not only in terms of data processing but also in optimizing the delivery of healthcare services. The fog layer, situated at the edge of the network, becomes an intelligent intermediary capable of pre-processing and filtering data before sending it to the cloud for more in-depth analysis. This hierarchical approach ensures that only relevant and critical information is transmitted, reducing bandwidth usage and enhancing the overall efficiency of data transmission and analysis.

The integration of fog computing with cloud computing represents a paradigm shift in healthcare services, providing a holistic and responsive solution to the industry's evolving needs. Fog computing's ability to address the challenges of latency, data security, and real-time analytics makes it a key enabler in elevating the quality of patient care. As healthcare continues to embrace digital transformation, the collaboration between cloud and fog computing stands out as an innovative approach that not only meets the demands of today's healthcare landscape but also paves the way for future advancements in medical technology and service delivery.

Fog computing is a transformative technology that complements cloud computing, particularly in the context of healthcare. While cloud computing offers numerous advantages, it also presents certain limitations, including issues related to latency, bandwidth, and data privacy. Fog computing addresses these challenges by bringing data processing closer to the source, which has profound implications for the healthcare sector.

5.3.1 Reduced latency

In healthcare, real-time data processing is often mission-critical. Whether in telemedicine consultations, remote patient monitoring, or surgical procedures, minimizing latency is paramount to ensuring the timeliness and effectiveness of healthcare interventions. Fog computing plays a pivotal role in achieving low-latency data processing.

Local Data Processing: Fog computing minimizes latency by processing data locally, closer to where it is generated, rather than transmitting it to distant cloud servers. This approach ensures that critical data is analyzed and acted upon swiftly, mitigating delays and enabling near-instantaneous decision-making. For example, in telemedicine applications, when a physician needs to assess a patient's vital signs in real time, fog computing can analyze information at the edge of the network, ensuring that the physician receives immediate feedback, which is crucial for making accurate diagnoses and treatment recommendations.

5.3.2 Enhanced data privacy

Data privacy and security are paramount concerns in healthcare, given the sensitive nature of patient information. Fog computing contributes significantly to enhancing data privacy by enabling localized data processing and analysis.

Reduced Data Transmission: Unlike traditional cloud computing, which often requires the transmission of sensitive patient data to centralized cloud servers for processing, fog computing allows for the analysis of data at network's edge or within local healthcare facilities. This means that patient information, including medical records and diagnostic data, remains within the confines of the healthcare facility, with lower risk of breaches of data or unauthenticated access.

Regulatory Compliance: The localized approach of fog computing aligns well with regulatory requirements such as the Health Insurance Portability and Accountability Act in the USA. Healthcare organizations must perform to strict data privacy regulations, and fog computing helps them meet these obligations by minimizing the exposure of patient data to external networks.

5.3.3 Edge devices in healthcare

The proliferation of medical devices and wearables has transformed patient care by providing continuous monitoring and real-time health data. Fog computing empowers these edge devices to process data locally, which has several benefits for healthcare applications. See Figure 5.5, which shows data privacy and security are paramount concerns in healthcare.

Local Intelligence: Medical devices, such as insulin pumps, cardiac monitors, and wearable fitness trackers, often require real-time decision-making. Fog computing enables these devices to analyze data locally and make intelligent decisions without the need for constant communication with a distant cloud server. For instance, an insulin pump can continuously monitor glucose levels and adjust insulin doses in real time, enhancing patient safety and treatment accuracy.

Reduced Network Congestion: By offloading data processing tasks to edge devices, fog computing reduces the strain on network bandwidth and cloud servers. This is particularly critical in healthcare settings with multiple connected devices, ensuring that critical data can be transmitted without delays or bottleneck.

5.4 LITERATURE REVIEW

Abdullah and Al-Kateeb (2022) delve into the concept of integrated healthcare with the Internet of Things (IoT) and assess its current status. They provide a systematic review of the technology's placement in both home and hospital settings, along with a comprehensive classification of challenges and issues hindering its progress.

Andriopoulou, Dagiuklas, and Orphanoudakis (2017) discuss the migration of computing intelligence from the cloud to the edge network, emphasizing the advantages of fog computing in healthcare with its proximity to users and low response times. The chapter introduces an architectural model and employs use cases to showcase the integration of IoT and fog computing.

Da Silva and Sofia's (2020) paper discusses the concept of context-awareness and examines various types of context-awareness indicators utilized in edge selection algorithms. The paper encompasses current approaches, the algorithms' roles, their scope, and the performance metrics considered.

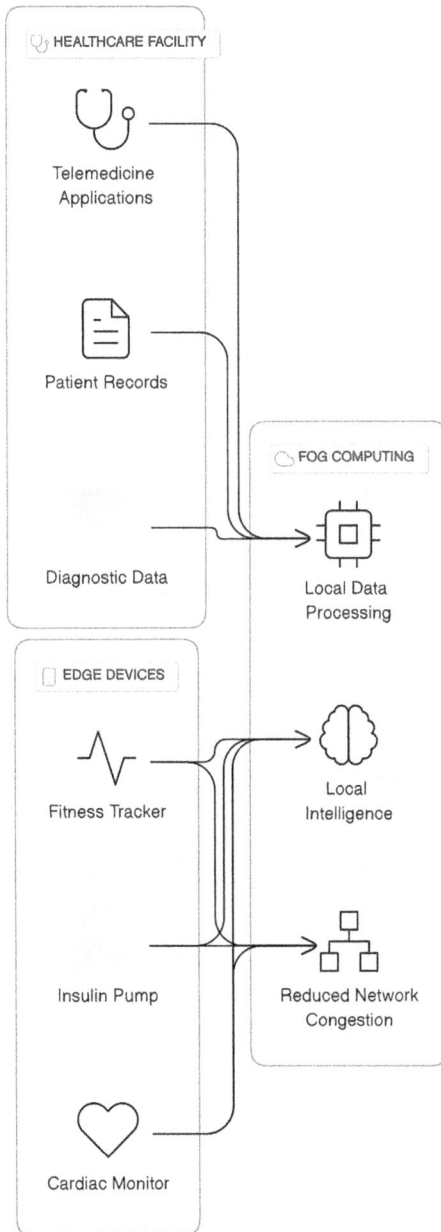

Figure 5.5 Data privacy and security are paramount concerns in healthcare.

Doukas and Maglogiannis (2012) present a cloud-based platform for managing mobile and wearable healthcare sensors, showcasing the application of the IoT paradigm in pervasive healthcare.

Gowda et al. (2022) conducted research aiming to enhance the quality of healthcare services through the integration of IoT and fog computing. The focus is on optimizing one or more key parameters to improve healthcare delivery.

Harnal et al. (2023) conducted a comprehensive analysis focusing on emerging cloud-based trends and applications within the healthcare sector. Their study also examined the challenges and obstacles encountered in the healthcare discipline related to cloud technology adoption.

Hassanalieragh et al. (2015) present an analysis that underscores both the opportunities and challenges associated with IoT in realizing the envisioned future of healthcare.

Kim et al. (2010) introduce the chord for cloud (C4C) scheme, which reduces the number of authentication requests sent to the identity provider and distributes the authentication process within the federated cloud environment using the chord algorithm.

Rahmani et al. (2018) present a smart e-health gateway that leverages edge network gateways for local storage, real-time data processing, embedded data mining, and other advanced services in healthcare.

Shi et al. (2015) provide an examination of the characteristics of fog computing and the potential services it can offer within the healthcare system.

Wang & Jin (2019) paper presents a comprehensive overview of the latest optimization methods for mobile cloud computing (MCC). These methods are designed to address diverse priorities and achieve optimal trade-offs among multiple objectives in MCC scenarios.

Ye et al. (2022) introduced an innovative and efficient public auditing scheme tailored for cloud-assisted health IIoT systems.

5.5 CHALLENGES AND CONSIDERATIONS IN INTEGRATING CLOUD COMPUTING WITH FOG COMPUTING IN HEALTHCARE

While the integration of cloud computing with fog computing presents a promising solution for elevating healthcare services, it is not without its challenges and considerations. One of the primary hurdles lies in the complexity of managing a hybrid architecture that involves both cloud and fog computing. Healthcare organizations must navigate the intricacies of coordinating data processing tasks between the centralized cloud and the distributed fog layer. This requires a careful evaluation of the types of tasks suitable for each layer, striking a balance between leveraging the computational power of the cloud for intensive analytics and utilizing fog computing for real-time, low-latency applications.

Data security and privacy emerge as critical considerations in the integration of cloud and fog computing in healthcare. While fog computing minimizes data transmission to the cloud by processing information at the edge, it introduces additional points of potential vulnerability. The decentralized nature of fog computing means that security measures must be implemented at various edge devices, introducing a more complex security landscape. Healthcare organizations need robust strategies to ensure the confidentiality and integrity of patient data at every point in the hybrid architecture. Encryption, secure authentication, and continuous monitoring become imperative to safeguard against potential security breaches.

Interoperability and standardization pose substantial challenges in the integration of cloud and fog computing within healthcare systems. Different healthcare devices, sensors, and applications may operate on disparate standards, hindering seamless communication between the cloud and fog layers. Establishing common standards for data exchange and communication protocols is essential to ensure interoperability, allowing healthcare systems to function cohesively and share information effectively. The lack of standardized practices can lead to integration difficulties, increased development costs, and potential delays in the deployment of integrated cloud-fog solutions.

Reliability and resilience are paramount in healthcare services, and the integration of cloud and fog computing introduces additional points of potential failure. The distributed nature of fog computing means that edge devices may be susceptible to various environmental factors, hardware malfunctions, or network disruptions. Healthcare organizations must implement redundancy and failover mechanisms to ensure continuous service availability. This involves developing strategies to seamlessly switch between cloud and fog resources in case of failures, minimizing downtime, and maintaining the reliability of healthcare services.

Scalability is a consideration that becomes particularly significant as healthcare data continues to grow exponentially. While cloud computing offers scalable resources, the integration of fog computing requires careful planning to accommodate the increasing volume of data at the edge. Healthcare organizations must anticipate the scalability requirements of both the cloud and fog layers to ensure that the infrastructure can handle the expanding demands of data-intensive applications, such as real-time monitoring and analytics.

While the integration of cloud computing with fog computing holds immense potential for elevating healthcare services, addressing the challenges and considerations is paramount for successful implementation. Healthcare organizations need to carefully navigate the complexities of managing a hybrid architecture, prioritize data security and privacy, establish interoperability standards, ensure reliability, and plan for scalability. Overcoming these challenges will pave the way for a more resilient, responsive, and efficient healthcare ecosystem, where the integration of cloud and fog computing optimally contributes to improved patient outcomes and healthcare delivery.

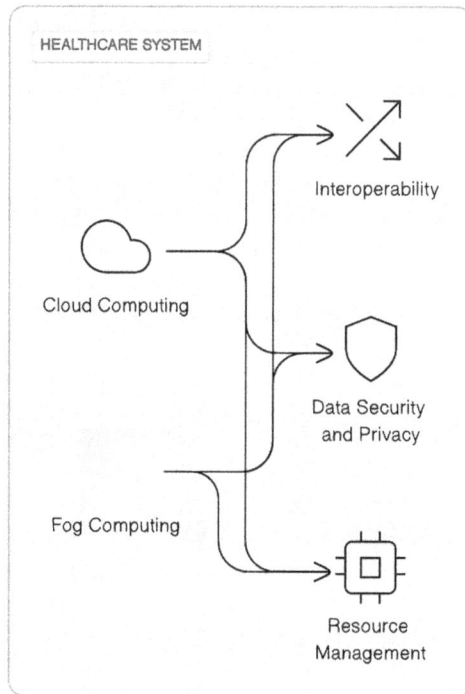

Figure 5.6 Integration of cloud and fog computing in healthcare.

The fusion of cloud and fog computing within the healthcare sector signifies a potential paradigm shift. However, it brings along intricate challenges and factors that demand meticulous attention to guarantee its prosperous deployment and functioning. See Figure 5.6, which shows integration of cloud and fog computing in healthcare.

5.5.1 Data privacy and data security

Data Security: Ensuring security of patient information is paramount. Healthcare organizations must implement robust security measures to protect data, whether it is stored in the cloud or processed at the edge. Data encryption, access controls, intrusion detection systems, and security audits are needed segments of comprehensive security.

Privacy Concerns: Healthcare data is highly sensitive, containing personal and medical information that must be safeguarded. Compliance with data-privacy regulations, such as HIPAA in the United States, is essential to prevent legal and ethical issues. Healthcare providers must be vigilant in protecting patient confidentiality and getting consent for data gathering and analyzing.

Cloud vs. Edge Security: Distinct security challenges exist for cloud and fog computing. While cloud providers typically offer robust security measures, edge devices may have varying levels of security. Healthcare organizations must ensure that edge devices are protected against physical and cyber threats to maintain data integrity.

5.5.2 Interoperability

Legacy Systems: Many healthcare organizations are still based on legacy systems, which cannot be easily compatible with cloud and fog computing platforms. Achieving seamless communication and data sharing between these diverse systems can be intricate. Standardized protocols, interoperability standards, and middleware solutions are necessary to guarantee compatibility.

Data Exchange: Effective data exchange between cloud and fog components is vital for delivering cohesive healthcare services. Standards like Fast Healthcare Interoperability Resources (FHIR) and Health Level7 (HL7) are critical for ensuring that patient data flows seamlessly between cloud and edge devices, regardless of the underlying systems.

5.6 RESOURCE MANAGEMENT

Resource management is a critical aspect of elevating healthcare services, and the integration of cloud computing with fog computing emerges as a transformative solution in this domain. Cloud computing has revolutionized healthcare by offering scalable, on-demand access to a plethora of resources, enabling healthcare providers to store, process, and analyze vast amounts of patient data efficiently. However, the conventional cloud model faces challenges, particularly in healthcare scenarios where low-latency and real-time data processing are imperative. This is where fog computing comes into play. Fog computing extends the capabilities of the cloud closer to the edge of the network, reducing latency and improving response times. By integrating cloud and fog computing, healthcare organizations can optimize resource allocation, ensuring that critical data is processed locally for immediate decision-making, while less time-sensitive data is offloaded to the cloud for more in-depth analysis. This integration enhances the overall efficiency of healthcare services, particularly in scenarios such as remote patient monitoring, where timely intervention is crucial. Moreover, the hybrid cloud-fog model enhances data security and privacy by minimizing the transmission of sensitive information to the cloud.

The synergy between cloud and fog computing in healthcare resource management extends beyond data processing. It also involves the efficient utilization of computational resources, storage, and networking infrastructure. Cloud resources can be leveraged for heavy computational tasks, such as complex analytics and machine learning algorithms applied to large data sets. Meanwhile, fog computing facilitates the processing of data from

Internet of Things (IoT) devices, wearables, and medical sensors at the network edge. This distributed approach reduces the burden on centralized cloud servers and optimizes the utilization of computing resources. Furthermore, by incorporating artificial intelligence (AI) algorithms into the fog layer, healthcare systems can gain insights from real-time data, facilitating predictive analytics for personalized patient care.

The integration of cloud and fog computing in healthcare resource management not only enhances the efficiency of healthcare services but also contributes to cost-effectiveness. Cloud computing offers economies of scale, allowing healthcare providers to access computing resources on a pay-as-you-go basis. Meanwhile, fog computing minimizes the need for continuous high-bandwidth connections to the cloud, reducing data transfer costs and alleviating network congestion. This cost-effective approach enables healthcare organizations to allocate resources judiciously, focusing on delivering improved patient outcomes without compromising on data security or real-time responsiveness. In conclusion, the integration of cloud computing with fog computing represents a paradigm shift in healthcare resource management, offering a holistic solution that addresses the unique challenges of the healthcare landscape while fostering innovation, efficiency, and cost-effectiveness in the delivery of healthcare services.

5.7 FUTURE PROSPECTS OF CC AND FC COMPUTING FOR HEALTHCARE

The future prospects of integrating cloud computing (CC) with fog computing (FC) in healthcare herald a transformative era for elevating healthcare services to unprecedented levels of efficiency, accessibility, and innovation. As technology continues to advance, this integration is poised to play a pivotal role in shaping the future landscape of healthcare delivery. One of the key aspects that will drive these prospects is the evolution of edge devices and IoT technologies. The proliferation of wearable devices, medical sensors, and smart healthcare devices will generate an ever-increasing volume of real-time patient data. The marriage of cloud and fog computing allows for the seamless processing of this data, combining the vast computational power of the cloud with the low-latency processing capabilities of the fog, ensuring that healthcare professionals have timely access to critical information for making informed decisions.

Artificial intelligence (AI) and machine learning (ML) will also be instrumental in the future of cloud and fog computing in healthcare. The integration of these technologies at both the cloud and fog layers enables advanced analytics, predictive modeling, and personalized medicine. The cloud's high computational power is harnessed for training complex AI models on extensive data sets, while fog computing facilitates the deployment of these models at the edge for real-time decision support. This AI-driven approach will revolutionize diagnostics, treatment plans, and patient care, leading to more precise and personalized healthcare interventions.

Telemedicine and remote patient monitoring are areas where the integration of cloud and fog computing is expected to make significant strides in the future. The ability to process and analyze patient data at the edge, near the point of care, ensures that remote monitoring applications can provide timely insights without relying solely on a centralized cloud infrastructure. This is especially crucial for scenarios such as continuous monitoring of chronic conditions, post-operative care, and emergency response systems. As healthcare services become more decentralized and patient-centric, the cloud-fog integration will enable a seamless and responsive healthcare experience, transcending geographical boundaries.

Another exciting prospect lies in the development of 5G networks, which will play a pivotal role in enhancing the capabilities of cloud and fog computing in healthcare. The high data transfer speeds and low latency offered by 5G networks will further optimize the real-time processing of healthcare data, enabling applications that demand instantaneous responses. This is particularly significant in emergency situations, where split-second decisions can be life-saving. The integration of 5G with cloud and fog computing will foster a more connected and responsive healthcare ecosystem, supporting applications such as telesurgery, augmented reality (AR) for medical training, and remote consultations with unprecedented clarity and immediacy.

Ethical considerations and regulatory frameworks will also shape the future prospects of cloud and fog computing in healthcare. As patient data becomes more distributed between the cloud and edge devices, ensuring robust privacy and security measures will be paramount. Regulatory bodies will need to adapt and establish guidelines that govern the ethical use of AI in healthcare, patient consent, and data protection. Collaboration between technology developers, healthcare providers, and regulatory agencies will be crucial to building a framework that fosters innovation while safeguarding patient rights and confidentiality.

In conclusion, the integration of cloud computing with fog computing represents a promising future for healthcare services. As technology continues to advance, the synergy between these two paradigms will drive unprecedented improvements in healthcare delivery, from real-time patient monitoring to personalized medicine. The future holds a healthcare landscape where cloud and fog computing work harmoniously to create a responsive, efficient, and patient-centric ecosystem, ultimately contributing to enhanced patient outcomes and an evolution in the way healthcare services are delivered and experienced.

The integration of CC with FC in healthcare is a dynamic and evolving field that holds immense potential for shaping the future of healthcare delivery. See Figure 5.7, which shows future aspects of fog computing and cloud computing in healthcare.

Several exciting developments are on the horizon.

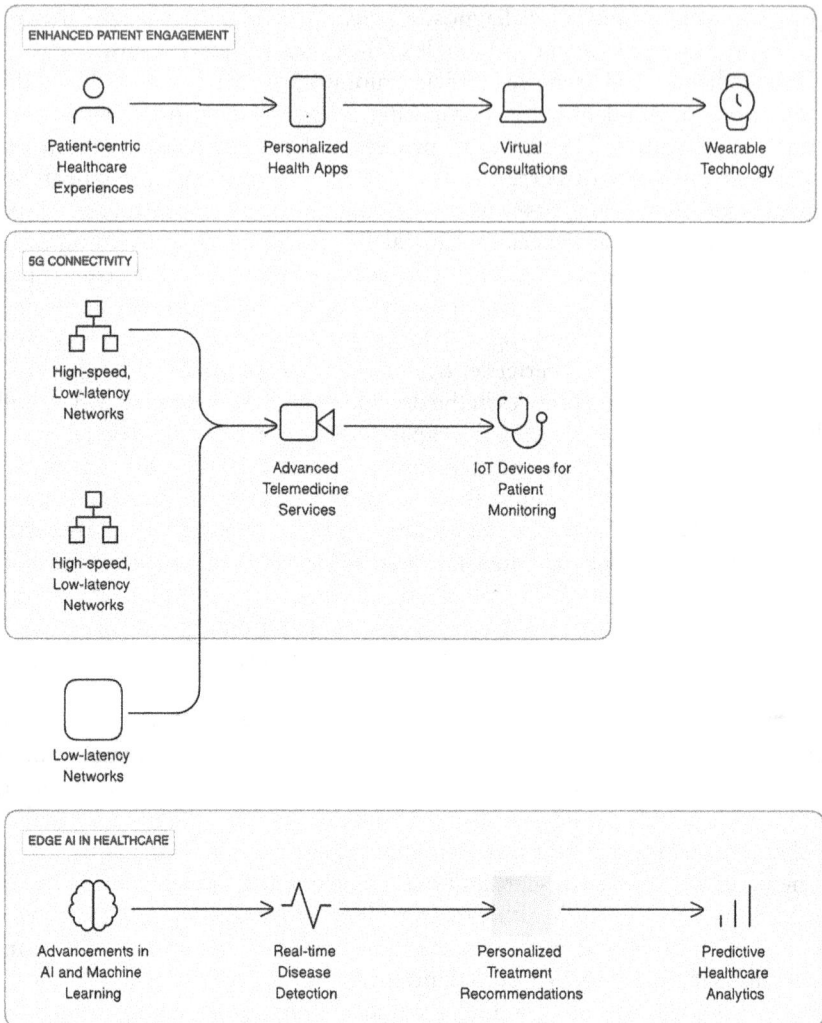

Figure 5.7 Future aspects of fog computing and cloud computing in healthcare.

5.7.1 Edge AI in healthcare

Advancements in AI and Machine Learning: As AI with ML algorithms continues to advance, their integration with edge computing in healthcare will become more prevalent. Edge AI presents the power of intelligent decision-making closer to the source of data generation, which is especially valuable in healthcare.

Real-Time Disease Detection: Edge AI will enable real-time disease detection and diagnosis. For example, wearable devices with AI could regularly monitor critical signs and detect anomalies, such as irregular heart rhythms or changes in glucose levels, providing early warnings to healthcare providers and patients.

Personalized Treatment Recommendations: Edge AI can analyze patient data at the point of care to provide personalized treatment recommendations. This could include optimizing medication dosages based on real-time data or tailoring treatment plans to a patient's unique genetic profile.

Predictive Healthcare Analytics: Fog computing, with its capacity for real-time data analysis, will facilitate predictive healthcare analytics. AI-driven models can predict disease outbreaks, patient readmissions, and treatment responses, allowing healthcare providers to proactively address healthcare challenges and allocate resources effectively.

5.7.2 5G connectivity

High-speed, low-latency networks: Deployment of 5G networks is poised to significantly amplify the capabilities of fog computing in healthcare. These high-speed, low-latency connections will provide the infrastructure required for advanced telemedicine services and the widespread adoption of IoT devices for patient monitoring.

Advanced Telemedicine Services: 5G connectivity will enable high-quality, real-time video consultations with healthcare providers, even in remote areas. This will expand access to care and support telemedicine applications like remote diagnosis, tele-surgery, and tele-rehabilitation.

IoT Devices for Patient Monitoring: 5G networks will facilitate the use of a vast array of IoT devices for patient monitoring. These devices can continuously transmit patient data to fog computing nodes, where real-time analysis can trigger alerts or interventions as needed.

5.7.3 Enhanced patient engagement

Patient-centric Healthcare Experiences: Cloud and fog computing will pave the way for more patient-centric healthcare experiences. Patients will have greater access to their health data and more control over their healthcare decisions.

Personalized Health Apps: Cloud and fog computing will enable the development of personalized health applications that offer actionable insights, reminders for medication, and health recommendations tailored to individual patients.

Virtual Consultations: Patients will be able to schedule virtual consultations with healthcare providers, eliminating geographical barriers and enhancing convenience. These consultations can include real-time video appointments and secure messaging for medical queries.

Wearable Technology: Wearable devices will continue to evolve, offering features like continuous health monitoring, early warning systems, and data sharing with healthcare providers. Patients can actively engage in managing their health and share relevant data seamlessly with their healthcare team.

5.8 CONCLUSION

In conclusion, the integration of cloud computing with fog computing emerges as a transformative force in elevating healthcare services, marking a paradigm shift in the way healthcare is delivered, managed, and experienced. This synergistic approach addresses the unique challenges of the healthcare landscape, offering a dynamic solution that combines the strengths of centralized cloud infrastructure with the agility of edge computing. The seamless integration of these technologies optimizes resource management, enhances data security, and ensures real-time responsiveness, thereby significantly improving patient outcomes and healthcare efficiency. The collaborative power of cloud and fog computing is particularly evident in applications such as remote patient monitoring, AI-driven diagnostics, and telemedicine, where the low-latency processing capabilities of fog computing complement the vast computational resources of the cloud. Moreover, the future prospects of this integration, driven by advancements in edge devices, AI, 5G networks, and ethical considerations, promise to revolutionize healthcare further, creating a patient-centric, connected ecosystem that transcends geographical barriers. As the healthcare industry continues to embrace digital transformation, the integration of cloud with fog computing stands at the forefront, not just as a technological innovation but as a catalyst for a holistic and responsive healthcare experience. The journey toward elevating healthcare services through this integration is marked by ongoing advancements, interdisciplinary collaboration, and a commitment to leveraging cutting-edge technologies for the betterment of patient care on a global scale. It is within this transformative landscape that the true potential of cloud and fog computing in healthcare unfolds, reshaping the narrative of healthcare delivery and charting a course toward a more connected, efficient, and patient-focused future. In other words, the integration of fog computing with cloud computing is reshaping the healthcare landscape. By combining the scalability of cloud computing with the low latency of fog computing, healthcare vendors may give more efficient, secure, and patient-centered care. As technology continues to advance, the healthcare industry must adapt and embrace these innovations to provide the highest standard of care to patients worldwide. This integration represents a pivotal moment in the evolution of healthcare, where the fusion of digital technology and medical expertise is poised to revolutionize the industry in unprecedented ways.

REFERENCES

Abdullah, D. B., & Al-Kateeb, Z. N. (2022, August). Prospects and challenges of the cloud of things in telemedicine. In 2022 8th International Conference on Contemporary Information Technology and Mathematics (ICCITM) (pp. 1–7). IEEE. DOI: 10.1109/ICCITM56309.2022.10031829

Andriopoulou, F., Dagiuklas, T., & Orphanoudakis, T. (2017). Integrating IoT and fog computing for healthcare service delivery. Components and services for IoT platforms: Paving the way for IoT standards, 213–232. DOI: 10.1007/978-3-319-42304-3_11

Da Silva, D. M. A., & Sofia, R. C. (2020). A discussion on context-awareness to better support the IoT cloud/edge continuum. IEEE Access, 8, 193686–193694. DOI: 10.1109/ACCESS.2020.3032388

Doukas, C., & Maglogiannis, I. (2012, July). Bringing IoT and cloud computing towards pervasive healthcare. In 2012 Sixth International Conference on Innovative Mobile and Internet Services in Ubiquitous Computing (pp. 922–926). IEEE. DOI: 10.1109/IMIS.2012.26

Gowda, D., Sharma, A., Rao, B. K., Shankar, R., Sarma, P., Chaturvedi, A., & Hussain, N. (2022). Industrial quality healthcare services using Internet of Things and fog computing approach. Measurement: Sensors, 24, 100517. DOI: 10.1016/j.measen.2022.100517

Harnal, S., Sharma, G., Malik, S., Kaur, G., Simaiya, S., Khurana, S., & Bagga, D. (2023). Current and future trends of cloud-based solutions for healthcare. Image Based Computing for Food and Health Analytics: Requirements, Challenges, Solutions and Practices: IBCFHA, 115–136. DOI: 10.1007/978-3-031-22959-6_7

Hassanalieragh, M., Page, A., Soyata, T., Sharma, G., Aktas, M., Mateos, G., ... & Andreescu, S. (2015, June). Health monitoring and management using Internet-of-Things (IoT) sensing with cloud-based processing: Opportunities and challenges. In 2015 IEEE international conference on services computing (pp. 285–292). IEEE. DOI: 10.1109/SCC.2015.47

Kim, I. K., Pervez, Z., Khattak, A. M., & Lee, S. (2010). Chord based identity management for e-healthcare cloud applications. 2010 10th IEEE. In IPSJ International Symposium on Applications and the Internet. DOI: 10.1109/SAINT.2010.68

Rahmani, A. M., Gia, T. N., Negash, B., Anzanpour, A., Azimi, I., Jiang, M., & Liljeberg, P. (2018). Exploiting smart e-Health gateways at the edge of healthcare Internet-of-Things: A fog computing approach. Future Generation Computer Systems, 78, 641–658. DOI: 10.1016/j.future.2017.02.014

Shi, Y., Ding, G., Wang, H., Roman, H. E., & Lu, S. (2015, May). The fog computing service for healthcare. In 2015 2nd International Symposium on Future Information and Communication Technologies for Ubiquitous HealthCare (Ubi-HealthTech) (pp. 1–5). IEEE. DOI: 10.1109/Ubi-HealthTech.2015.7203325

Wang, X., & Jin, Z. (2019). An overview of mobile cloud computing for pervasive healthcare. IEEE Access, 7, 66774–66791. DOI: 10.1109/ACCESS.2019.2917701

Ye, W., Wang, J., Tian, H., & Quan, H. (2022). Public auditing for real-time medical sensor data in cloud-assisted health IIoT system. Frontiers of Optoelectronics, 15(1), 29. DOI: 10.1007/s12200-022-00028-1

Chapter 6

IoT to IoE

Fundamentals, layered architecture, application areas security principles, challenges, and their remediation

Himanshi Agrawal and Neeti Bansal

Meerut Institute of Engineering & Technology, Meerut, Uttar Pradesh, India

6.1 IoT Vs IoE: Fundamentals

Information and communication technology has seen tremendous growth over the past few decades with the emergence of IoT. People are switching from traditional table desktops to smartphones and tablets and smart household devices that can communicate their data over the internet. IoT has connected everything to everyone. It creates a physical environment where people are connected with their things and have the power to remotely monitor and control them. IoT comprises a large number of connected devices (things) that can sense and collect data and transmit it over networks. This data is then stored, analyzed, and converted into meaningful information that can be used to drive IoT application.

IoE on the other hand provides a broader view by integrating things, processes, people, and data connected through a continuous intelligent network. Also, IoE is considered as an amalgamation of machine to machine (M2M), people to people (P2P), and machine to people communication (M2P) [1]. Figure 6.1 (Source: Cisco IBSG, 2012) [2] depicts this collaboration, which leads to an intelligent ecosystem called IoE.

Therefore, unlike IoT, which comprises only intelligent devices (things), IoE involves people with processes, things, and data to improve quality and standard of living [3].

CISCO predicted a tremendous growth of this technology in 2013, with $14.4 trillion expenditure on collaboration of M2M, M2P, and P2P communication for implementing IoE [4].

Both terms, "IoT" and "IoE", can be more clearly understood from the example of a smart home equipped with intelligent appliances, security cameras, and smart lighting systems. IoT works on connecting and controlling these devices individually via internet. The user can access the data of various devices on a smartphone and can control them. However, IoE goes a step ahead, where the intelligent appliances, security cameras, and smart lighting systems are interconnected among themselves as well as

DOI: 10.1201/9781003461418-6

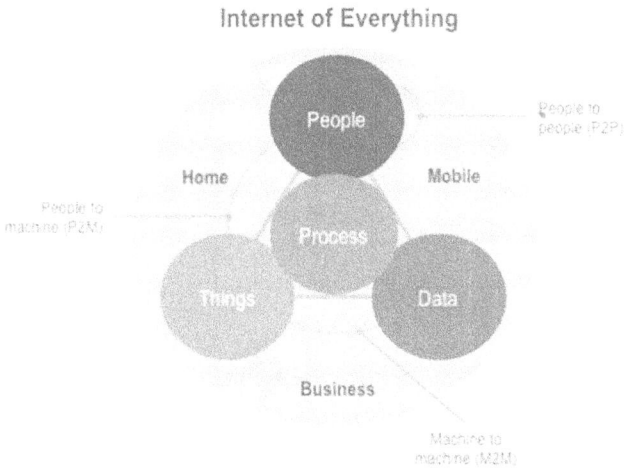

Figure 6.1 IoE components.
Source: [Cisco IBSG, 2012] [2].

to the owner's smartphone. Such a system serves multiple tasks of remote monitoring and control, automation, security, energy usage optimization, etc., resulting in a better quality of life.

6.2 THE EMERGENCE OF IoE

The evolution to IoE is driven by a fundamental shift in perspective. Instead of just focusing on devices and data, IoE encompasses a more holistic view of interconnected systems, including not only devices but also people, processes, and data. The key differences between IoT and IoE can be summarized as follows.

Scope: IoT primarily involves the connection of physical objects, whereas IoE extends beyond things to encompass people, organizations, and processes. It aims to create an all-encompassing digital ecosystem.

Data integration: While IoT often results in data silos, IoE emphasizes data integration and analytics across multiple domains. It promotes a unified view of data to extract meaningful insights and drive innovation.

Interactions: In IoT, interactions primarily occur between devices and systems. In contrast, IoE emphasizes human-to-machine, machine-to-machine, and human-to-human interactions, creating a more immersive and interconnected experience.

Intelligence: IoE leverages advanced technologies like artificial intelligence (AI), machine learning, and cognitive computing to make sense of the vast amounts of data generated. It enables systems to learn, adapt, and predict outcomes autonomously.

6.3 IoT vs. IoE: Layered architecture

The previous section suggests that IoT and IoE are related terms, with IoE presenting a broader vision. This implies that the layered architecture for these two must be different. The layers that define IoT cannot be used to define the functionality of IoE as it includes additional roles and responsibilities.

The five-layered architecture defined for IoT comprises a perception layer, transport layer, processing layer or middleware, application layer, and business layer [5–8]. The functionality of the different layers is discussed as follows.

Perception layer: It comprises intelligent IoT-enabled devices or things that are embedded with sensors and actuators. This layer is thus capable of collecting data from the environment and performing specific actions.

Network layer: It provides the data communication from the perception layer to other higher layers.

Middleware: This layer, also known as the data processing layer, provides the important functionality of data storage, processing, and analysis with the help of databases, cloud computing platforms, and big data/data analytic tools.

Application layer: This layer processes the results of middleware and delivers value and services to the end user.

Business layer: It is the top layer that controls the entire IoT application, defines business and profit modules, and ensures end-user privacy.

IoE layers will be defined based on IoT layered architecture, but they must also include the layers required to implement the added functionality offered by the technology. The first two layers (perception and network layer) will remain the same as in IoT. The third layer must be a data collection/storage layer, followed by a processing layer, data abstraction layer, application layer, and business layer. Here, the functionality of the first two layers will remain the same. However, the functionality of the remaining layers will differ, as shown in Figure 6.2.

As shown in the figure, the first two layers of IoT and IoE have same functionality. However, the middleware in IoT is divided into three different layers, including a data collection/storage layer, processing layer, and data abstraction layer.

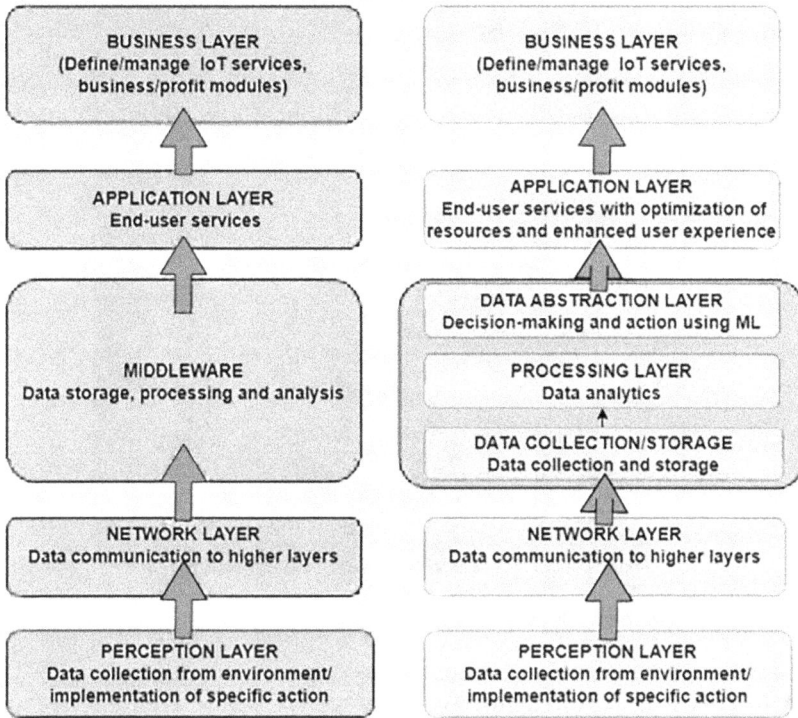

Figure 6.2 IoT and IoE layered architecture.

Data collection/storage layer: This layer collects data from the lower layers (from intelligent devices and things) and from other resources such as business and enterprises and social media.

Processing layer: It applies data analytic tools on collected data to draw meaningful information.

Data abstraction layer: It applies machine learning (ML) techniques to form decisions on the information provided by the processing layer.

The application layer of IoE will provide end-user services with an objective of optimum utilization of resources like power and enhanced user experience, leading to a better living standard.

The resulting seven-layered architecture for IoE resembles the seven-layered architecture given by IoT World Forum (IoTWF) committee in collaboration with Cisco, IBM, etc. in 2014 to extend the capabilities of IoT [9,10].

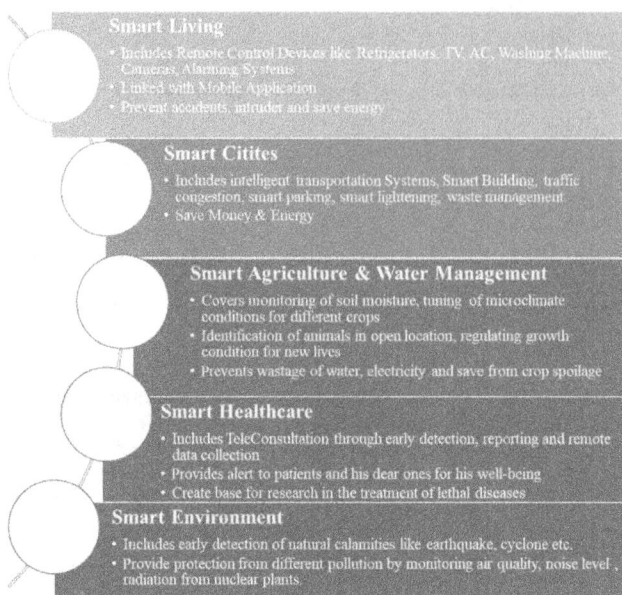

Figure 6.3 IoT and IoE application areas.

6.4 IoT vs IoE: Application areas

IoT and IoE both support and provide assistance in all sense to people and various organizations. Figure 6.3 categorizes the areas that utilize IoT and IoE.

Figure 6.3 provides an overview of applications of IoT and IoE. Both of these technologies are spreading their horizons at a fast pace. They are providing a new way to run any industry to make operations digitized and realizing the goals of Industry 4.0.

Both IoT and IoE are involved in widespread areas across health, environment, agriculture, defense, manufacturing units, etc.

While IoT provides interconnection of devices and provides their remote monitoring and control, IoE moves a step ahead by connecting people to people, machines to machines, and people to machines, resulting in enhanced user experience, better quality of life, and optimization of available resources.

6.5 SECURITY PRINCIPLES, CHALLENGES, AND THEIR REMEDIATION

The IoT and IoE have revolutionized the way we interact with the digital world. These interconnected networks of devices and sensors have permeated every aspect of our lives, from smart homes and healthcare to transportation and industrial automation. However, this pervasive connectivity also brings forth a multitude of security challenges that must be addressed to ensure the safety and privacy of users and the integrity of the systems. In this chapter, we will explore the security principles, challenges, and their remediation in the realm of IoT and IoE.

The IoT and IoE landscapes offer immense opportunities for innovation and connectivity. However, the complexity and scale of these ecosystems present significant security challenges that cannot be ignored. By adhering to security principles, addressing challenges, and implementing remediation strategies, we can build a safer and more secure IoT and IoE environment that protects data, privacy, and infrastructure while harnessing the potential of these technologies for a better-connected world.

6.5.1 Security principles

6.5.1.1 End-to-end encryption

Ensuring data confidentiality and integrity is fundamental. Employing end-to-end encryption ensures that data remains secure throughout its journey from the sensor or device to the cloud or other endpoints.

6.5.1.2 Device authentication

Authenticating devices before granting them access to the network prevents unauthorized devices from joining, reducing the risk of breaches and data manipulation.

6.5.1.3 Data integrity

Data can be tampered with during transmission or storage. Implementing mechanisms like hash functions and digital signatures can verify data integrity and authenticity.

6.5.1.4 Access control

Role-based access control (RBAC) should be enforced to limit the privileges of devices and users. This principle reduces the attack surface and minimizes the potential damage from a security breach.

6.5.1.5 Regular updates and patch management

Regularly updating the firmware and software of IoT devices is crucial to fix known vulnerabilities and ensure the latest security features are in place (Figure 6.4).

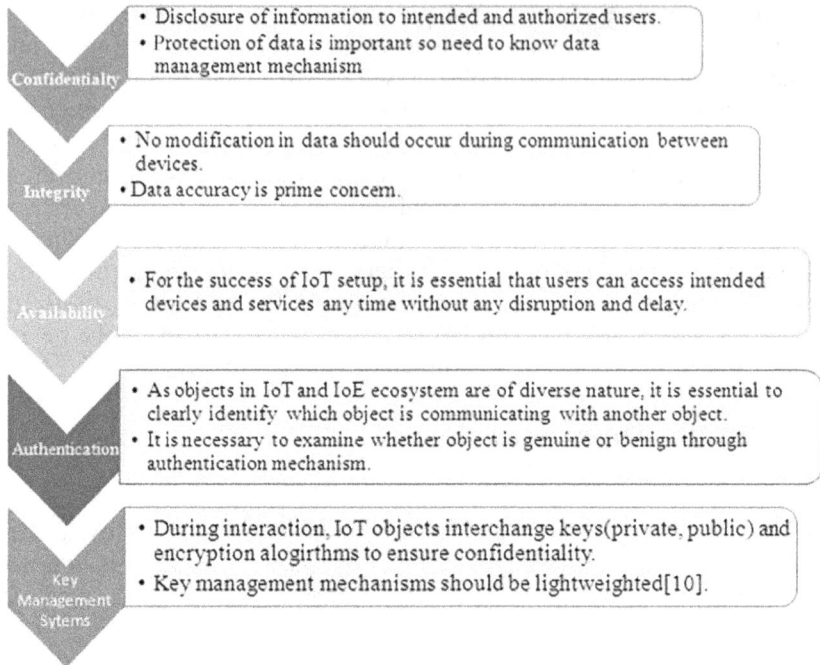

Confidentialty
- Disclosure of information to intended and authorized users.
- Protection of data is important so need to know data management mechanism

Integrity
- No modification in data should occur during communication between devices.
- Data accuracy is prime concern.

Availability
- For the success of IoT setup, it is essential that users can access intended devices and services any time without any disruption and delay.

Authentication
- As objects in IoT and IoE ecosystem are of diverse nature, it is essential to clearly identify which object is communicating with another object.
- It is necessary to examine whether object is genuine or benign through authentication mechanism.

Key Management Sytems
- During interaction, IoT objects interchange keys(private, public) and encryption alogirthms to ensure confidentiality.
- Key management mechanisms should be lightweighted[10].

Figure 6.4 Security principles.

6.5.2 Security challenges

6.5.2.1 Resource constraints

Many IoT devices have limited processing power and memory, making it challenging to implement robust security measures without compromising performance.

6.5.2.2 Diversity of devices

The IoT ecosystem comprises devices from various manufacturers with different operating systems and security capabilities. Ensuring uniform security across this diversity is difficult.

6.5.2.3 Data privacy

Collecting vast amounts of data from IoT devices can raise concerns about user privacy. Unauthorized access to this data can have serious consequences.

6.5.2.4 Scalability

As the number of IoT devices continues to grow exponentially, managing and securing a large-scale deployment becomes increasingly complex.

6.5.2.5 Physical vulnerabilities

IoT devices deployed in the field are often physically accessible to potential attackers, making them susceptible to tampering or theft.

6.5.3 Remediation strategies

6.5.3.1 Hardware-based security

Utilize hardware security modules (HSMs) and trusted platform modules (TPMs) to enhance device security at the hardware level, making it more resistant to physical attacks.

6.5.3.2 Machine learning and AI

Implement AI and machine learning algorithms to detect anomalies in device behavior, which can help identify potential security threats in real time.

6.5.3.3 Security standards and certification

Adhere to industry-specific security standards and obtain certifications to ensure devices meet established security criteria.

6.5.3.4 Edge computing

Process data at the edge of the network to reduce the volume of data transmitted, minimizing the attack surface and improving real-time security.

6.5.3.5 Blockchain technology

Implement blockchain for secure and transparent data storage and transaction verification, enhancing data integrity and trust.

6.5.3.6 User education

Educate users about the importance of security best practices, such as strong password management and regular updates, to mitigate human-related security risks.

6.6 FUTURE OF IoE

The evolution from IoT to IoE represents a paradigm shift in the way we view and interact with technology. It promises a future where the digital world is seamlessly integrated into our physical lives, enabling new levels of convenience, efficiency, and innovation. However, it also presents challenges related to privacy, security, and ethical considerations that must be carefully navigated.

As IoE continues to mature, it is essential for businesses, governments, and individuals to embrace this transformative technology responsibly, ensuring that it benefits society as a whole while safeguarding against potential risks. The journey from IoT to IoE is one of continuous innovation and adaptation, and its full potential is yet to be realized. The only certainty is that it will continue to shape the future in ways we can only begin to imagine. As we enter this new era of the Internet of Everything, it's not just about connecting things; it's about connecting everything, including ourselves, to a smarter and more interconnected world.

REFERENCES

1. A. Raj, and S. Prakash, "Internet of Everything: A Survey Based on Architecture, Issues and Challenges," *2018 5th IEEE Uttar Pradesh Sect. Int. Conf. Electr. Electron. Comput. Eng. UPCON 2018*, Dec. 2018, doi: 10.1109/UPCON.2018.8596923.
2. D. Evans, "The Internet of Everything How More Relevant and Valuable Connections Will Change the World," 2012.
3. M. H. Miraz, M. Ali, P. S. Excell, and R. Picking, "A Review on Internet of Things (IoT), Internet of Everything (IoE) and Internet of Nano Things (IoNT)," *2015 Internet Technol. Appl. ITA 2015—Proc. 6th Int. Conf.*, pp. 219–224, Nov. 2015, doi: 10.1109/ITECHA.2015.7317398.
4. A. N. J. Bradley, J. Loucks, and J. Macaulay, "Internet of Everything (IoE) Value Index: How Much Value Are Private-Sector Firms Capturing from IoE in 2013?," *Cisco Internet Bus. Solut. Gr. (IBSG), Cisco Syst. Inc., San Jose, CA, USA, White Pap. 2013*, 2013.
5. I. Mashal, O. Alsaryrah, T. Y. Chung, C. Z. Yang, W. H. Kuo, and D. P. Agrawal, "Choices for Interaction with Things on Internet and Underlying Issues," *Ad Hoc Networks*, vol. 28, pp. 68–90, May 2015, doi: 10.1016/J.ADHOC.2014.12.006.
6. R. Khan, S. U. Khan, R. Zaheer, and S. Khan, "Future Internet: The Internet of Things Architecture, Possible Applications and Key Challenges," in *Proceedings – 10th International Conference on Frontiers of Information Technology, FIT 2012*, 2012.
7. E. Cavalcante, M. P. Alves, T. Batista, F. C. Delicato, and P. F. Pires, "An Analysis of Reference Architectures for the Internet of Things," in *Proceedings of the 1st International Workshop on Exploring Component-based Techniques for Constructing Reference Architectures – CobRA '15*, 2015, pp. 13–16, doi: 10.1145/2755567.2755569.

8. A. Al-Fuqaha, M. Guizani, M. Mohammadi, M. Aledhari, and M. Ayyash, "Internet of Things: A Survey on Enabling Technologies, Protocols, and Applications," *IEEE Commun. Surv. Tutorials*, vol. 17, no. 4, pp. 2347–2376, 2015, doi: 10.1109/COMST.2015.2444095.

9. D. Hanes, G. Salgueiro, P. Grossetete, R. Barton, and J. Henry, *IoT Fundamentals: Networking Technologies, Protocols, and Use Cases for the Internet of Things*. Cisco Press, 2017.

10. R. Mahmoud, T. Yousuf, F. Aloul, and I. Zualkernan, "Internet of Things (IoT) Security: Current Status, Challenges and Prospective Measures," IEEE, in *the Proceeding of the 10th International Conference for Internet Technology and Secured Transactions (ICITST)*, 2015, doi: 10.1109/ICITST.2015.7412116.

Chapter 7

Issues in managing multimedia big data

Shailendra Prakash[1], K. Rama Krishna[2], and Ajay Kumar Gupta[1]

[1]Associate Professor, Department of Computer Science & Engineering, IIMT College of Engineering, Greater Noida, Uttar Pradesh, India
[2]Professor, MCA Department, GNIOT Engineering Institute, Greater Noida, Uttar Pradesh, India

7.1 INTRODUCTION

Big data is a composite concept with diverse/changeable explanations [1]. Big data has five basic characteristics: capacity (huge quantity of information produced, usually from diverse causes), assortment (planned or amorphous information), rate (range of statistics given), inconsistency (frequently altering data), and genuineness (statistics). The ability to excavate and scrutinize these statistics effectively can divulge formerly indefinite comprehension, trends, and associations that can be utilized to aid resolution creation [2,3]. Due to the amalgamation of data sets, content can frequently change, and data can be streamed from different origins.

A large amount of (big) data generated originates from Internet of Things (IoT) devices and this volume increases with the efficiency of IoT devices [4]. The quantity of data generated by IoT devices reached 600 zetta bytes (ZB) per year in 2020 [5]. The expansion of multimedia big data (MMBD) is escalating every day with the help of the internet and mobile technologies. Consumer expends an ample of moments connecting with every one to distribute the facts through the internet [6].

Users may have little networking expertise, but they may be able to forward and access the consumer-formed stuff (together with transcript, speech, picture, and record) through the internet to a societal arrangement [6]. The massive multimedia stuff on societal networks encloses a great extent of multimedia information from diverse resources, from recorded surveillance to social media.

MMBD has an enormous quantity of statistics in a haphazard manner, along with digital relevance/function settings, like digital recovery, consent, etc. [4]. To address authentic globe troubles, multimedia examination practitioners deliberate on the concerns of organizing, managing, exhuming, construing, and skillfully analyzing various types of data. Yet, diverse explorers and technical endeavors have diverse metaphors for this term. Enormously huge statistics accumulation is also recognized as huge statistics.

DOI: 10.1201/9781003461418-7

Furthermore, by Apache Hadoop, huge facts are described as "data sets which can not be collected, controlled, and processed within an appropriate scope by general systems" [5]. MMBD representation is very essential as MMBD is indefinite, multi-model, assorted, and has an elevated level of difficulty in details. Furthermore, these high-level semantics cannot be understood by the machine. Multimedia information is a category of data sets that is human-centric, varied, and larger than traditional big data. However, compared to MMBD, traditional big data requires fewer complex algorithms and less computational resources. Visibility in media framework, information, individual-centered, elevated, and semantics is associated with MMBD [6].

Multimedia huge facts can be viewed as significant indications collected of principally mysterious composite configuration that includes a trail of motion, instance information, spatial figures, personal aspects, and organized singularity association [1]. Multimedia huge information is also amorphous, has numerous molds, and is assorted, which is tough to signify and mold.

Multimedia huge statistics have the requirements of quality of knowledge (QOK), which also progress with instance and space. Huge statistics in multimedia also require significantly extra assets in terms of collection, saving, broadcast, management, and dispensation, integrating, for illustration, the necessity for graphics dispensation unit (GDU) dispensation and disseminated analogous soft skills [5].

In this chapter, the authors will converse the tribulations that arise in MMBD and address them through existing security frameworks. The instigator will first address the MMBD and concentrate on MMBD protection troubles, and then explain the need for protection and current methods that are used to protect MMBD. After that, we will wrap up the manuscript to accomplish the enhanced outcome like exhilarating explores promotion and superior industry conclusion; organizations require realizing it efficiently.

7.1.1 Life cycle of multimedia big data

The emblematic multimedia huge information life cycle is exemplified in Figure 7.1, which embraces acquiring, condensing, accumulating, processing, accepting, estimating, calculating, and protecting. In the acquirement step, multimedia big data is frequently produced by assorted resources, counting abundant convenient mobile devices (like Android phones, digital-cams, personage digital gadgets, etc.), recorded speech, digital playoffs, internet, IoT, inter-media feeler, societal medium, and fundamental vocabulary. After acquirement multimedia, huge statistics were compressed for its proficient saving and broadcasting, since its capacity is beyond the capability of common software/association utensils. In the saving and dispensing steps, one countenances the issue of unmatched statistics amount.

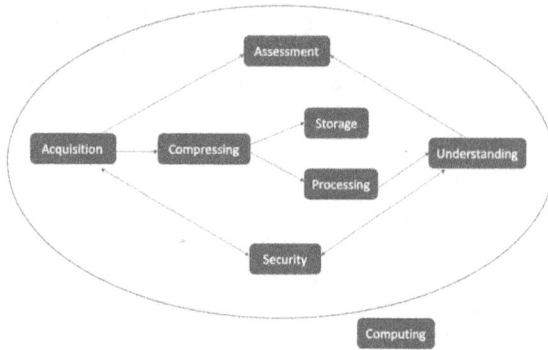

Figure 7.1 Biorhythm of MMBD.

Besides, it is repeatedly infeasible to accumulate all confined multimedia information in numerous consequences. Current multimedia huge statistics examination schemes are desirable to accomplish speedy storage and handing out. Since there is a semantic fissure between semantics and recorded illustration facade, in voluntary thought it is requisite for multimedia huge statistics when a consumer obtains multimedia big data. To recover consumers' requirements for inter-media huge facts jobs, we necessitate appraising the advantages of multimedia huge statistics. With the speedy development of susceptible statistics, the rudiments multimedia big facts are to administer the huge information under noise protection throughout the complete multimedia huge facts biorhythm.

In the typical multimedia big data environment, bio-rhythm calculation yields insightful outcomes, emphasizing the importance of authentic real-time data and efficient analysis at every stage of the multimedia big data lifecycle. Given the massive amounts of data and the need for timely progress, accurate analysis is crucial for the advancement of multimedia big data functions within an acceptable timeframe.

7.1.2 Features of MMBD

Multimedia big data processing necessitate far more complex algorithms and a larger computational power than the conventional huge statistics. An essential characteristic of multimedia big data is the efficient handling of media frameworks and content, along with the provision of high-level, human-centered insights. Here are some typical qualities of multimedia big data.

In contrast to non-multimedia huge statistics, multimedia huge statistics comprises an additional statistics category, which is extra sympathetic for individual accepting than for machinery (for illustration, the population is often familiar with tape more than a transcript, but it is complicated for an instrument to comprehend tape rather than transcript). From media category to media stuff, multimedia huge information emphasizes people, not machinery.

Facts arrays are chiefly collected from diverse forms of videotape statistics like cam-tape, interactive videotape, societal videotape, immersive videotape (also entitled stereoscopic 3D videotape), 3D digital global, and so on. Therefore, multimedia huge statistics have a more complicated stage of involvement than typical huge statistics, which is usually transcript-dependent.

Inter-media huge information can be as assimilated from assorted resources, counting pervasive transferable mobile apparatus (such as the Android phone, D-cam, person age digital appliance, etc.), tape speech, D-playoffs, the internet, the IoT, inter-media feeler, societal media, and fundamental language. Diverse resources compose multimedia huge facts to be amorphous, unrelated, and numerous molds. Therefore, it is incredibly thought-provoking to signify and mold multimedia huge statistics since this informational approach from diverse resources or positions (for illustration, societal, virtual, and corporeal). Meanwhile, a measurable fraction of multimedia huge statistics is mislaid and multimedia huge fact sheets are often imperfect due to probable dissimilar genesis.

Considering the constant reaction of multimedia huge information over instance and area, multimedia huge statistics investigate desires to meritoriously consider the framework and stuff in diverse circumstances.

Inter-media huge information has unrivaled information size, in which numerous purposes knock into huge, dynamic, composite statistics rivulet and concern to unconventional systematic skills. Furthermore, it is regularly inappropriate to accumulate all imprison multimedia statistics in various cases.

Since multimedia statistics is broadcasted through a rapid network, multimedia big statistics desires to be managed speedily and incessantly, subject to the storage amount and moment constraint. Real-time computation is compulsory for accumulated huge facts in multimedia because the statistics are also huge to convey. Furthermore, multimedia huge facts are characteristically active, i.e., transitory in the accepted globe, and also illustrate an ample multiplicity since it regularly transforms with instance and space.

Amid the production of susceptible tape statistics, the necessity of multimedia big data requires it to be managed securely. It is hazardous to believe collaborator tracing and protected inter-media contribution concerns for multimedia huge information.

7.1.3 Importance and applications of big data

The time is not too far when large statistics will be turned into a prime source for expansion of individual firms. In view of increasing competition and the ability to detain the relevant information, all companionship will have to acquire huge statistics critically [7]. This will also lay the foundation for balancing huge fact activities, like large infrastructure development, platform progress, and procedures in integration complex and fact-driven troubles in disciplines and engineering. This is very useful for the operation of information aspects and the formation of new facets for decision-makers.

Recently, numerous U.S. organizations, like the NIH (National Institute of Health) and the NSF (National Science foundation), have realized that the use of information-intensive management would have a valuable effect on their potential expansion. So, they are attempting to make large- scale data technology and techniques to smooth out their assignment through the U.S. government's large-scale big data initiative [7]. As per the McKinsey Institute report [8], the competent use of huge information has hidden payback to alter the financial system and give a novel gesture for dynamic growth.

The utilization of invaluable water resources, coupled with extensive data, will emerge as a primary challenge in today's endeavors. This challenge may lead to the emergence of new competitors capable of attracting skilled employees through significant data-driven insights. Explorers, policy-makers, and verdict producers have to distinguish the capability to use large information to highlight the subsequent gesture of development in their locale. There are numerous paybacks in the industry sector, which can be obtained by big data, to augment equipped effectiveness; inform strategically; enlarge enhanced client examine; discover and develop novel goods and services; novel clientele and identity, etc. Some applications of big data are explained as follows.

7.1.3.1 Real-time analysis

Previously, if business users lacked the technical proficiencies required for large analysis, they had to inquire their IT contemporaries for assistance. Generally, when they access the demanded data, it was no longer practical or accurate. With the largest autistics utensils, the scientific group can work with rapid algorithms to fetch the information. In other words, they can enlarge the organization and mount interactive and energetic visibility utilities that permit the business consumer to scrutinize observation and construct revenue from facts [9–13].

7.1.3.2 Personalized healthcare

We inhabit an increasingly diverse world, yet the healthcare sector often continues to adopt a generalized approach in certain areas. When someone is diagnosed with cancer, they usually go through medical cure, and if that does not work, then the doctor tries other procedures, etc. But what if a cancer patient can get medicine as per his genes? This will provide better results, lower costs, fewer nuisances, and less horror. With personal genome planning and large facts, this will soon be ordinary because everybody has a medical record. Hence, by finding genetic determinates, personalized medicine can be given to patients to treat the disease and its causes [14].

7.1.3.3 Data security

An organization's data can be mapped with the help of big data tools, which allows searching and analyzing the threats that the organization might face internally. As an outcome, the organization will be able to identify insecure information that is potentially sensitive, and also it will be able to check whether it is stored according to the regulatory requirements or not [9].

Big data technology is pending as a prime IT constituent for the majority of organizations and industries and there's an emergent focus on the verdict of the business payback of huge information analytics functions to facilitate to rationalize the savings in them. Huge information is being used in various services and has a vast range of applications in the different domains, as exposed in Figure 7.2. Big data leads to big-time benefits for searchers and organizations. Web- based applications often encounter large amounts of data, such as community calculating, counting societal networking investigation non-line interactions, referral structure, standing structure, and forecasting marketplace internet transcript and credentials, internet investigation guide, and other recent hot spots. There are immeasurable sensors around us that produce small amounts of sense figures that are required to be utilized, such as intelligent transportation systems (ITSs), depending on the analysis of a huge quantity of combined sensor statistics. Extensive e-commerce is intensive because it engrosses a huge amount of client data and, hence, there exists a demand for it in the IT world [8,15].

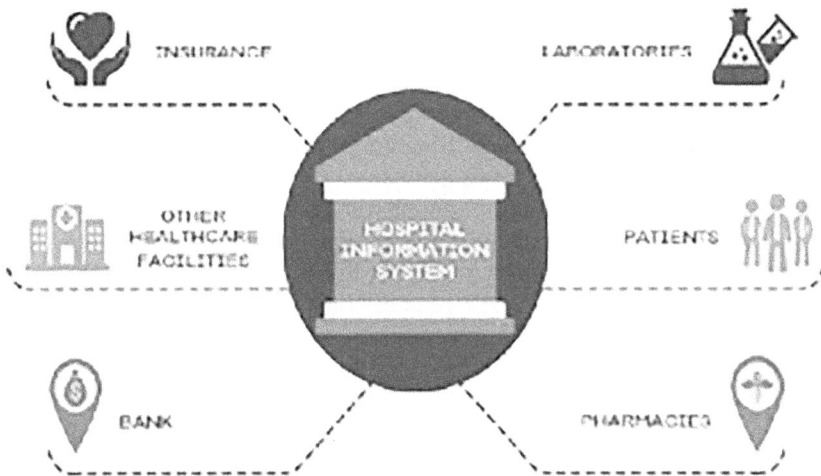

Figure 7.2 Applications of MMBD.

7.2 SAFETY CORRELATED MULTIMEDIA BIG DATA TRIBULATIONS, CHALLENGES, AND THREATS

The big data analyzers' biggest problem is how to diminish processing time and storage space while maintaining the findings from the tiny statistics sets as reliable as those. Parallel processing is a way to execute a task in a distributed environment via numerous figuring assets at the alike instance and is a solo prime way in assigning an unbeaten cogent. To attain this, several large statistics analytics proposals have been proposed, consisting of IBM big data analytics [1], MS Azure [8], Oracle huge statistics analytics [16], etc., which explore statistics in an effective manner.

In the modern large statistics era, innovative prospects and confrontations emerge with elevated-diversity multimedia statistics jointly with the gigantic extent of societal statistics. Hence, multimedia huge statistics cogent has engrossed a large concentration in mutual academe and commerce in current time. It is measured as a rising and demanding matter due to its decisive and priceless insight [2,3]. The concern and confrontation of multimedia big statistics protection are versatile and can be seen with diverse aspects.

7.2.1 Transportability

When it is seen from the discernment of transportability, the payback encompasses the availability and effortless accessibility. When customers admit statistics from the diverse position, the admission rights may be modified as of the transform in the positions [4,6]. For an illustration, suppose a physician attempts to admit a medicinal description. He might acquire them from the interior of the hospital arrangement or from the exterior association like their home. So he/she might be permissible to those medicinal descriptions when admittance from the restricted association of the hospice and not while the person is off the job. So the spatio-temporal confines have to be in use while setting up the admittance management and it ought to end in an unswerving manner.

Take the situation of liability administration decision; for the consumer to understand the strategy in an improved way [1,8,16], it should not to be prepared assorted. A consumer might not be an intellectual to utilize them and may get be besieged with it. To acquire the highest gain from it, it ought to be position-aware, which offers true instance and applicable information. Additionally, it should provide the necessities associated with confidentiality that embrace administrative facts, assemble the statistics, explain the intent, and tapping. Administration signifies the admittance and usage of inter-media information, while assembly describes the statistics compilation in a definite style in accordance with the protection strategies. Explanation of intent treaty by the difficulty of how and by whom the information is admittance. Tapping engrosses tapping of the facts concerning individuals who accumulate and utilize the facts.

7.2.2 Multi-level fortification

Presently, various defense molds have survived, with prominence largely on organizer-level fortification. Except in the situation of inter-media statistics, it is not appropriate since typically inter-media includes numerous figures of essentials considering the diverse stage of "sensitivity". Observation tape normally stresses an elevated rank defense as compared to a few standard multimedia statistics. So, the method to proffer safety ought to be supple and it must manage admittance on several stages, which consents to the background of admittance authorizations for the diverse category of inter-media items in an inter-media flow [17].

7.2.3 Huge statistics may accumulate from various ending peaks

The information in huge statistics arrives from a wide range of resources consisting of customers of societal networks, customers of cell phones, and the users of the Internet of Things. It is noticed that with a thorough statistics review, statistics coming from various resources are sought to be incorporated [1,16].

7.2.4 Data accumulation and distribution

Data accumulation and allocation ought to be securely contained by the context of a prescribed, reasonable framework: The convenience of information and lucidity of its existing and history of use by statistics consumers is an imperative façade of huge information. In a few situations, wherever configurations are misplaced or collected haphazardly, there may be an inevitability of communal or protected garden gateway and ombudsman, like roles for statistics at relaxing. These organization amalgamations and unanticipated amalgamations call for enhanced huge facts construction [2].

7.2.5 Information hunt and assortment necessitate confidentiality and recaution guidelines

There is a deficiency of methodical sympathetic of the competency that should be supplied by a statistics supplier in this respect. An amalgamation of erudite consumers, erudite engineers, and organization defenses may be desirable as well as exclusive of record or preventive inquiry that may be predicted as facilitating re-recognition. The main attribute of big data is, as an analyst said, that the ability to develop important observations from advanced analytics on data at any selection, the track and assortment feature of analytics will ensure protection and privacy concerns [3,18], privacy conserve utensils are desired for big data, such as for individually identifiable information (III). Because there may be dissimilar, potentially unpredicted dispensation steps among the statistics owner, supplier, and facts consumer, the confidentiality and uprightness of facts coming from finish summit should be protected at all points [17].

7.3 REQUIREMENTS OF MULTIMEDIA BIG DATA PROTECTION

The extent of multimedia big statistics protection is massive as is the massive quantity of inter- media facts that are produced all day. Societal media and films, examination recorded-tapping, consumer multimedia statistics, etc. are each in the coercion of protection procedures and strategies in position to guard their statistics from illicit admittance. There is a necessity to incorporate various admittance control strategies. Huge statistics are incorporated statistics from dissimilar causes. Huge statistics possess their admittance to organize strategy entitled "sticky strategy". So whenever they are absorbed with statistics by various foundations, their innovative strategy ought to be imposed. As there is a necessity to integrate the approaches and proffer a solution to the clashes concerned [19].

Authorizations must be robotically imposed, predominantly for allowance permissions. It is not practical to physically manage admittance controls on huge statistics as it desires a good granule admittance management [1]. Therefore, a few procedures are mandatory to robotically permit agreement. Managing multimedia data admittance manages a composite strategy. This is the nature of admittance to organize, which is a crucial solitary where the sanction is approved or deprived of depends on the inter-media statistics stuff [8,20]. This category of admittance organization is tremendously considerable when trading with statistics, which desires protection, such as examination cam-shots. In the case of huge information sets of inter-media nature, this can be quite challenging to mind the content-based admittance organization, and to have understanding of the inter-media that necessitates to to being confined [21] in the controversy of admittance management strategies in huge statistics stores.

With some of the hot innovations in large information organizations, consumers are allowed to apply for impulsive jobs with programming languages such as Java in support. This generates numerous disputes which efficiently delay the execution of a satisfactory admittance control method for customers to automatically plan, create and execute the admittance organize strategy. In the case of environment that are energetic in nature where the sources, function, and users are continuously developing, it is mandatory to propose and execute admittance control policies to ensure unwavering facts accessibility and confidentiality [21,22].

7.4 DIVERSE TACTICS FOR MULTIMEDIA BIG DATA PROTECTION

7.4.1 Task administration

For this incline, consumer-definite strategies are created for a consumer crowd. It is a challenging assignment to raise confidential strategies that signify the preconception finalized by the consumer to attain their statistics. The imperative façade that ought to be precise now embraces the subsequent inquiry: where, how, and who. Who recognizes them? Where is it departing to be packed? How is it definite? To the initial inquiry, the solitary response is to provide supremacy for the customers themselves so that they oversee their statistics in the dispersion of the information to supplementary consumers in a protected and clandestine style. On the contrary, an individual can describe their strategies and the consumer will be supplied by the alternatives to endorse or abandon it based on their preference. For the inquiry on how it is definite, the solitary resolution is to convulsion them in admittance organize arrangement for effective supervision of confidential strategies within a venture. For the third difficulty of where it is stored, cloud tactics are a solution. Cloud tactics propose truthful hoardage in an effective mode that minds measurability. It helps reduce the cost of physical communications or facilitates data accumulation [17,23].

7.4.2 Locality depends on admittance management

Locality-depended admittance is existing to consumers in this scheme. Now, consumers are forced to distribute their existing vicinity to the admittance organizing structure. Afterward, comprehensive facts of the transportability of the consumer will be apprehended by the structure and evaluated above an instance. It could be a reason for a group's concern if there is a crevice in the protection of these facts [16,23,24].

7.4.3 Encipher

Enciphering rules can be utilized alongside a few permutation procedures on facts to elude the unofficial admittance of inter-media statistics.

7.4.3.1 Record encipher

Inter-media information might be encoded by symmetric key enciphering algorithm. AES is the best algorithm. But whenever inter-media information survives, it upholds extra valuation expenditure. Videocassette jumble and discriminatory record enciphering are dual practices that are utilized generally to attain videocassette enciphering [8].

In a selective enciphering organization mold, a numerous tributary inter-media organization is measured initially. For this organization, there is a patron, and the whole numeral of media rivulet is not in the instance. There are sinking roots and middle roots that progress accumulated information. Some restrictions of the organization are definite as pursuits. Restriction of roots in organization: Let p, q, b, v, e, x, and x, be nth measurement arrays. Information flow can be computed as array p = (p1 ... pnth), which signifies the nth diverse facts flow in instances. Stream replica can be computed as array q = (q1 ... qnth), and qi is an integer of replicas of statistics streams pi, and identified as the rear eqi patrons exhibit pi [x].

7.4.3.2 Image encipher

Many statistics enciphering algorithms are accessible presently. But mostly they are utilized for transcript information and may not be valid for information into imagery. Also, there are numerous shortcomings for these algorithms as they may be based on chief blocking in its progression. So, discriminatory image enciphering schemes are utilized to avoid the blocking. It does not encipher the whole picture, but does for the part of the picture and thus reduces the calculation [16,17].

7.4.3.3 Acoustic enciphering

The term "protected tone" refers to an identification method used in cryptology to enco de acoustic signals over various communication standards such as radio, cellphone, etc. The standard enciphering algo-rithms are computationally expensive in this condition too. One of the major vital categories of acoustic statistics is vocalizations when outlooking from a safety opinion. Vocalizations desire extensive protection contrasting another acoustic category like melody or another amusement form. The discriminating encipher scheme can be utilized in this scenario too to moderate the analytical complication [7,8,15,24].

7.4.3.4 Fingermarking

For this practice, a solitary identification will be embedded into every consumer's replicas. This can be excavated to assist in recognizing a crook in the condition of unlawful escape. This is a rate effectual process that can put off illicit admission of information. To defend information from unlawful admittance, usage, and reorganization after an endorsed admis-sion of a consumer, solitary consumer facts similar to fingermarks can be transferred to the inter-media information storage inside every consumer's replica to find out the unofficial consumers. The rooted statics ought to be imperceptible to the consumers. It develops into an extensive topic in public association to preserve digital inter-media matter. D-finger marking can notice the consumer who has modernized the inter-media material illicitly. Whenever an illegitimate replica is perceived, the fingermark can be excavated from it and utilizing a few algorithms, the traitor can be effortlessly acknowledged. A water-marking practice will be utilized to establish the fingermarking statistics in the inter-media stuff [14,15].

7.5 OUTLINE OF ACCESSIBLE ALGORITHM

By the argument of segment II, it offers outcomes that presently is no stationary algorithm for diverse inter-media documents in huge statistics. For statistics content protection problems in cloud analytics, the organization is trusting a RSA or DES algorithm. These algorithms are utilized for enciphering huge figures earlier to content, consumer confirmation practice, or configuration of a protected medium for information broadcast. The amalgam algorithm is the amalgamation of RSA and DES for confidential key enciphers [29]. To encode the immense equantity of statistics, amalgam techniques are utilized in which the statistics are enciphered by utilizing symmetric proposals (DES or AES) and the solution is forwarded by utilizing a heterogeneous scheme (RSA).

7.5.1 Data encipher standard algorithm

Block cipher is utilized by the data encipher standard (DES) algorithm. It encodes the information in slab extent of 8 bytes each. The solution extent is 7 bytes. One byte is discarded from the solution. The DES algorithm depends on changeover and transformation encompassing 16 rounds. In each round, key and data bits are moved, permuted, XORed, and forwarded by the 8S-Box. Firstly, 64 bits of plain text is specified to early permutation (IP). IP divides it into two portions: left side plaintext (LPT), right side plaintext (RPT). These mutual fractions pass 16 rounds and recombine, shown in Figure 7.3. The same method in conflicting arrangements is used for deciphering. 2DES and 3DES are similar practices except that they utilize two and three different keys respectively [9,10].

7.5.2 Blowfish algorithm

The blowfish algorithm includes slab code of an 8-byte block. Solution dimensions in this algorithm diverge from 4 bytes to 56 bytes. Two chief divisions of this algorithm are sub-solution creation and data encipher. Encipher has 16 encircling as utilized in DES. Every encircling embraces solution-based transformation and information-based replacement [9,10].

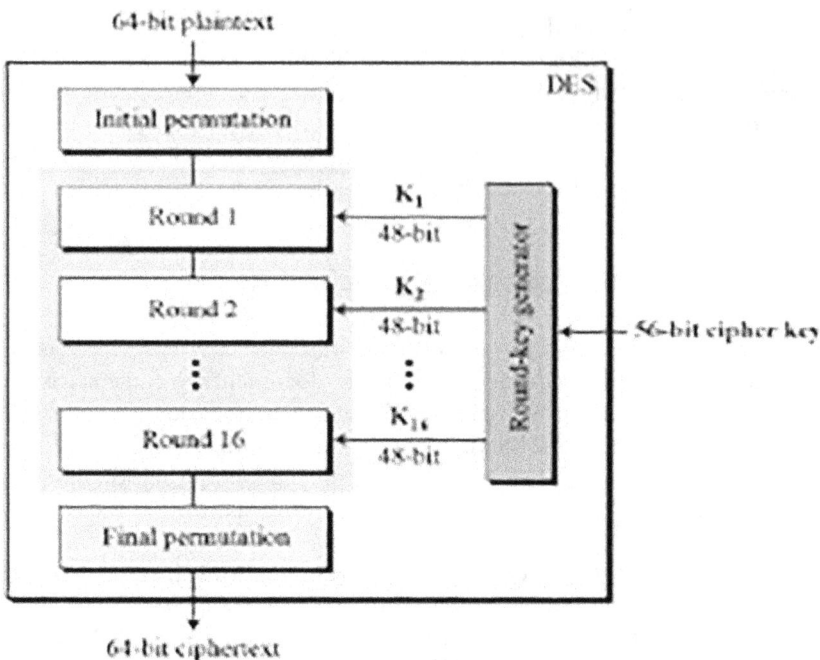

Figure 7.3 Rules of data encipher standard.

7.5.3 RSA algorithm

The RSA algorithm is a public-key enciphering algorithm enlarged by Ron Rivest, Adi Shamir, Len Adlemanin in 1977 [10]. It characteristically depends on an arithmetic perception. It utilizes prime figures to produce the public and private keys. This algorithm utilizes two dissimilar keys: solitary is a public key and next is the private key. Enciphering is completed by public key and decryption progression is completed by the private key.

- Utilizing an enciphering key (q,d), the algorithm is as follows.
- Represent the information as an integer amid 0 and (i-1). A prolonged memo can be disjointed into several blocks. Every block would then be represented by a numeral in the identical assortment.
- Encode the memo by lifting it to the qth power modulo d. The outcome is identical to the code text memo C.
- To decode code text memo C, lift it to another power K modulo d.
- The enciphering key (q,d) is prepared public. The decoding key (k,d) is reserved private by the consumer.

The question is how to decide suitable values for q, k, and d.

Prefer two very extensive (100+figure) prime figures. Signify these figures as t and u.

Placed identical tot*u
Prefer any outsized integer, k, like GCD (k, ((t-1)*(u-1)))=1
Find q such that q*k=1(modulus((t-1)*(u-1)))

7.5.4 Elevated measurement messy map

Acoustic records in huge statistics have complex forms and harsh jobs to administer and recommend protection. For acoustic records in multimedia, plentiful algorisms are utilized for enciphering and deciphering. Elevated measurement messy map utilizes an algorithm that includes low code algorithm that will be employed for enciphering information at a bit stage through a clandestine solution creator. It employs an elevated extent of solution store cover complications and augment potency alongside brute force similar to harassment. The purpose of this algorithm is quite reserved but the consumer can make it complex by escalating the key gap. The initial analog signal is transformed into a digital signal and then enciphering is attained on that digital indication [11].

7.5.5 H.264 video entropy ciphering

For record documentation with much information, H.264 is the inventive customary record ciphering. But there are a few protection disagreements in this algorithm. Presently, there are three classes of tape algorithms: complete enciphering algorithm, selective enciphering algorithm, and enciphering algorithm depending on entropy ciphering.

Absolute enciphering is enormously protected but it will hold abundant computational complications. Selective algorithms have fine issues, but they will modify the arithmetical recital of entropy and it will up shoot on video firmness quality. The third algorithm for video records is entropy ciphering, which has superior protection and firmness recital. H.264 is generally utilized for huge statistics in film records since it performs on the internet. H.264 contributes two entropy cipher schemes: context-adaptive variable length coding (CAVLC) and context-adaptive binary arithmetic coding (CABAC). The process of H.264 CAVLC is declared in the subsequent steps [12].

- Enciphers the entire numeral of all co-adjuvant and sprawling ones (total co-adjand trailing ones).
- Enciphers symbols of sprawling unity.
- Enciphers non-nilco-adjuvant points exclusive of trailing unity (point).
- Enciphers absolute numeral of zeros before final non-nilco-adjuvant (total zeros).
- Enciphers runs of nilco-adjuvant before every non-nilco-adjuvant (run before).

Here are a few boundaries of explained algorithms. It does not have the stipulation the novel H.265 firmness set. As knowledge departs, the narrative principles and layout are approached into the market, the protection must also update. This algorithm grants equilibrium among protection and ciphering obstacles. So, it is suitable for recorded communication, digital rights supervision, inter-media stock piles, etc. This algorithm does not illustrate fine recital on actual instance tape, which is on huge statistics.

7.5.6 DRM depends on enciphering

Digital rights management (DRM) is a conventional top preserve digitized storage tool with cryptology information. Identity-based enciphering generates simple record supervision through confidential information as the common solution. Secondly, the security of identity-based enciphering depends on the different record problem and uses the attributes of bilinear arrangement [13]. Identity-based enciphering constructs the solution for organization in an easy manner as well as also reducing the computation charge in comparison to conventional public solution cipher management [17]. DRM focused on the enciphering and will positively enlarge organization security, such as eaves drop assault, fake DRM section assault, and allocation assault, etc. For eaves drop assault, private properties are placed in the DRM section. So, the customer who duplicates the DRM drawing record can't analyze the solution even he replicates the encoded substances [26-28].

7.5.7 Elliptic curvature cryptology

Nealkoblitz and Victor Miller (1985) proffer the thought of utilizing elliptic curvature as a narrative category of cipher system, which focuses on being a substitute to RSA enciphering [25]. As the organization was mostly focused on elliptical curvature, it was identified by elliptic curvature cryptology (ECC). Cryptology is a digitized tactic that is utilized to protect the precious information for broadcast. Preserving our statistics by utilizing diverse validation practices is the chief purpose of cryptography. In the current era, ECC has a principal function in a source prescribed atmosphere as long as the key extent is not practicable in ad hoc wireless association and transportable networks. The foremost importance while embryonic ECC was to have an organization that utilizes an elevated stage defense public key cryptography organization. The public key organization includes two keys that are utilized for enciphering and deciphering progression. The initial solution is recognized as a communal solution that is generally scattered and subsequently is the individual or confidential solution which barely statistic predictable as a universal indication. The scheme of having a communal confidential solution was replicated competent schemes. ECC offers elevated altitude guard by dense dimension [25].

7.5.8 Tribulations related to ECC

The major intricacy with the elliptic curve cryptosystem is the effectiveness of the private key. A private key is an important step to produce a secured organization. This step cannot be ignored or the cryptosystem will be bridged inefficiently. Problems related with ECC include

- When K was stationary instead of haphazard, a cluster entitled fail over flow was proficient to convalesce the confidential solution.
- It is probable to repossess the confidential solution of a server via unlocking SSL that utilizes ECC against a digital figure by realizing a moment assault ("elliptic curve digital signature algorithm").
- Additional disagreement publicizes itself with elliptic curvature cryptology. The Java group protected haphazard little microorganisms in its execution that sometimes creates an impact in the worth of the confidential solution M=S*R. However, this trouble could be barred by the deterministic creation of S (elliptic curvature digital signature algorithm). This promises accurate unpredictability of S.

For the history of two decades, the National Security Agency (NSA) has been cheering ECC as an additional consistent replacement to RSA; but, in August 2015, the NSA blocked approval for the exploitation of ECC. This reason is that elliptic curvature cryptology is not protected [10,13].

7.6 PROPORTIONAL STUDY OF ALGORITHMS

Table 7.1 illustrates the proportional study among diverse symmetric and asymmetric algorithms at different environments of key algorithms like the key length, block length, rounds, strength utilization, avalanche result, dispensation instance, and numerous other platforms. Authors in [30] have concluded many contrasts among the algorithms of the equal category and achieved a conclusion that AES is quicker and more competent than all additional enciphering algorithms. In [31], the authors have programmed records with assorted storage and proportions. The result ensures that blowfish generate a finer code than the further enciphering algorithms and, as a result, the use of the blowfish algorithm was eminent. Blowfish has the maximum avalanche consequence due to the amount of exoraction, which modifies the efficiency drastically. DES has avalanche less than AES. RSA also has an elevated avalanche as it engrosses the arithmetical computation of two large prime figures. Now, in conversation about cryptanalysis conflict, the writer has clarified differential cryptanalysis for every algorithm. It was also observed that 3 DES and blowfish were vulnerable to brute-force assault, whereas in the condition of RSA, brute-force attack was difficult.

Table 7.1 A proportional examination of diverse cryptography algorithms

Algorithm	Year of Use	Key length	Size of block	No. of rounds	Power consumption	Avalanche effect
DES	1977	56-bits	64 bits	16	Low	Less than AES
AFS	2000	128 bits, 192 or 256 bit key	128 bits	10 (128 bits), 12 (192 bits), 14 (256 bits)	Low	Faster encryption/ decryptions less time than des
3DES	1978	168 bit, 112 bit or 56 bit	64 bits	48	Low as compared to DES, AES, blowfish, and RSA	Medium
BLOWFISH	1993	32 bits up to 448 bits	64 bits	16	high	Fastest except when changing keys
RSA	1977	>1024 bits	Min. 512 bits	No rounds	Very high	Slower encryption/ decryptions

Algorithm	Resources consumption	Security	Throughput	Cryptanalysis resistance	Tunability
DES	Requires more CPU cycle and memory	Inadequate	Medium	Vulnerable to linear and differential cryptanalysis	No
AES	Consumer resources when data and block size big	High	Very high	Strong against truncated differential, linear interpolation and square attack	No
JOES	Requires effective resource consumption	Vulnerable	Medium	Vulnerable to differential brute-force attacker can analyze plain text	No
10973 H	Requires pre-processing	High	High	Vulnerable to differential brute-force attacker	No
RSA	Very high	Very high	Very sigh	Brute-force attack to accomplish	Yes

AES was confirmed to be strong besides discrepancy, linear exclamation, and square assault [31]. Therefore, the split to AES algorithm has not been established yet. Contrasting with the further algorithms, DES is the mainly anxious algorithm as it has previously been declared sufficient to employ.

7.7 OPPORTUNITIES EXISTING IN MMBD

Besides every confront, the growing multimedia big data afford enormous approaches in human activities and emotions, which guide immense opening to making vast developments in many ways.

An elevated numeral of the portable consumer could supply statistics for distinguishing shopper activities and assist in the administrative procedure. It will also aid invention enhancement and ambition advertising. Appreciation to the sophisticated knowledge in statistics and wireless communication, movable illustration is one of the admired functions these days. By the speedy expansion of inter-media information, the claim for additional ability to explore and retrieve practice is more complex than ever in each case.

Analysis of any illness data can guide to the recognition of any illness at a premature stage and offer suitable treatment to prevent the possibility of getting infected. This information can also assist the population in awareness of healthy life. The industry could utilize large information to study more about their employees, enhance competence, and initiate innovative industry development [14].

The weather authorities also exploit large inter-media spatio-temporal statistics to study and examine the heat and weather structure of the planet. In addition, large statistics of astrophysics are able to learn the world extra-efficiently. This not only aids the investigator documentation and recognizes the world, but also helps to defend the planet from the shock of diverse space associations. The additional opening comprises the arrangement of tactical ways, the progress of superior client examines the recognition of novel goods and jobs, and so on [8].

7.8 CONCLUSION

The development of MMBD not only emerges with dares but also offers enormous openings for the expansion of multimedia huge data utility. In the multimedia huge statistics era, consumer necessities will stimulate the development of prospect multimedia knowledge. Multimedia huge statistics not only influence the person's living wages and thoughts, but also involves the community and financial expansion. This chapter highlights the study of different issues of multimedia big data and safety matter security tactics that are existing these days. A lot of investigation is being accepted on this ground of protection and a few strategies have been recognized to protect information from unlawful admittance and reorganization, such as position-based admittance organizing method and enciphering tactics. Therefore, we lastly wrap up our chapter with a vision that offers a comparative study among all existing techniques. We expect that the chapter will offer noteworthy apparition into the embryonic drift in the safety and confidentiality of multimedia big data.

REFERENCES

1. P. Zikopoulos, K. Parasuraman, T. Deutsch, J. Giles, D. Corrigan, and others, Harness the Power of Big Data The IBM Big Data Platform, McGraw-Hill Professional, New York, NY, 2012.
2. F. Fleites, H. Wang, and S.C. Chen, "TV shopping via multi-cue product detection", IEEE Transactions on Emerging Topics in Computing, vol. 3, pp. 161–171, 2015.
3. Y.F. Chang, C.S. Chen, and H. Zhou, "Smart phone for mobile commerce", Computer Standards & Interfaces, vol. 31, no. 4, pp. 740–747, 2009.
4. W. Zhu, P. Cui, Z. Wang, and G. Hua, "Multimedia big data computing", IEEE Multi Media, vol. 22, no. 3, pp. 96–106, Jul./Sep. 2015.
5. X. Wang, L. Gao, S. Mao, and S. Pandey, "CSI-based finger printing for indoor localization: A deep learning approach", IEEE Transactions on Vehicular Technology, vol. 66, no. 1, pp. 763–776, 2017.
6. Z.J. Wang, and S. Mao, "A survey of multimedia big data", Senior Member, IEEE, Lingyun Yang, and Pingping Tang.
7. X. Wang, L. Gao, and S. Mao, "CSI phase finger printing for indoor localization with a deep learning approach", IEEE Internet of Things Journal, vol. 3, no. 6, pp. 1113–1123, 2016.
8. J. Davidson, B. Liebald, J. Liu, P. Nandy, T.V. Vleet, U. Gargi, S. Gupta, Y. He, M. Lambert, B. Livingston, and others, "The YouTube video recommendation system", In Proceedings of the 4th ACM Conference on Recommender Systems. ACM, pp. 293–296, 2010.
9. R.L. Rivest, A. Shamir, and L. Adleman, "A method for obtaining digital signatures and public-key cryptosystems", ACM, vol. 21, no. 2, 120–126, Feb. 1978.
10. R. Gnanajeya Raman, and K. Prasadh Ramar, "Audio encipher using higher dimensional chaotic map", International Journal of Recent Trends in Engineering, vol. 1, no. 2, 2009.
11. C. Mian, J. Jia, and Y. Lei, "An H.264 video encipher algorism based on entropy coding", IEEE Xplore Digital Library, Article number = 4457649.
12. C.C. Yang, J.C. Hsiao, H.W. Yang, and J.Y. Jiang, "Robust DRM on internet based-on identity-based encipher", International Conference on New Trends in Information and Service Science, 2009.
13. Y. Lee, L. Batinaand, and I. Verbauwhede, "Untraceable RFID authentication protocols: Revision of EC-RAC", In Processing IEEE International Conference RFID, 2009.

14. R.M. Chezianand, and C. Bagya lakshmi, "A survey on cloud data security using encipher technique", International Journal of Advanced Research in Computer Engineering & Technology, vol. 1, no. 5, July 2012.
15. H. Xiong, "Structure-based learning in sampling, representation, and analysis for multimedia big data", In Proceedings of 2015 IEEE International Conference on Multimedia Big Data (Big MM), Beijing, China, pp. 24–27, 2015.
16. J. Bian, Y. Yang, and T.S. Chua, "Multimedia summarization for trending topics in microblogs", In Proceedings of the 22nd ACM International Conference on Information and Knowledge Management. ACM, pp. 1807–1812, 2013.
17. Z. Tufekci, "Big questions for social media big data: Representativeness, validity and other methodological pitfalls", In Proceedings of the Eighth International Conference on Web Logs and Social Media. Michigan, USA, pp. 505–514, 2014.
18. J.R. Smith, "Riding the multimedia big data wave", In Proceedings of the 36th International ACMSIGIR Conference on Research and Development in Information Retrieval. ACM, Dublin, Ireland, pp. 1–2, 2013.
19. C. Ye, Z. Xiong, Y. Ding, and K. Zhang, "Secure multimedia big data sharing in social networking using finger printing and encipher in the JPEG 2000 compressed domain", Intrust, Security and Privacy in Computing and Communication (TrustCom), 2014 IEEE 13th International Conference, 2014.
20. R. Toshiwal, K.G. Dastidar, and A. Nath, "Big data security issues and challenges", International Journal of Innovative Research in Advanced Engineering, vol. 2, 2015.
21. T.M. TriDo, J. Blom, and D.G. Perez, "Smart phone usage in the wild: A large-scale analysis of applications and context", In Proceedings of the 13th International Conference on Multimodal Interfaces, ACM, pp. 313–360, 2011.
22. M. Chen, "A hierarchical security model for multimedia big data", International Journal of Multimedia Data Engineering and Management, vol. 5, no. 1, pp. 1–13, 2014.
23. W. Zhu, P. Cui, Z. Wang, and G. Hua, "Multimedia big data computing", IEEE Multi Media, vol. 22, no. 3, pp. 96–106, 2015.
24. W. Tan, B. Yan, K. Li, and Q. Tian, "Image retargeting for preserving robust local feature: Application to mobile visual search", IEEE Transactions on Multimedia, vol. 18, no. 1, pp. 128–137, 2016.
25. S. Pouyanfar, Y. Yang, and S.C. Chen, "Multimedia big data analytics: A survey", ACM Computing Surveys, vol. 51, no. 1, 2018. doi: 10.1145/3150226
26. D. Hari Patil, and Dr. M. Singh, "A selective encipher and fraud detection on multimedia files for data security with resource optimization technique", vol. 5, no. 2, February 2018, JETIR (ISSN-2349-5162) JETIR 1802136.
27. M. Marjani, F. Nasaruddin, A. Gani, A. Karim, I.A.T. Hashem, A. Siddiqa, and I. Yaqoob, "Big IoT data analytics: Architecture, opportunities, and open research challenges", IEEE Access, vol. 5, pp. 5247–5261, 2017.
28. Liu, Z., Choo, K.K.R., & Zhao, M. , "Practical-oriented protocols for privacy-preserving out sourced big data analysis: Challenges and future research directions", Computers and Security, vol. 69, pp. 97–113, 2017.
29. R. Ali, C. Solis, M. Salehie, I. Omoronyia, B. Nuseibeh, and W. Maalej, "Social sensing: When users become monitors", In Proceedings of the 19th

ACMSIGSOFT Symposium and the 13th European Conference on Foundations of Software Engineering, ESEC/FSE' 11, pp. 476–479, 2011.

30. B. Sadiq, F.U. Rehman, A. Ahmad, M.A. Rahman, S. Ghani, A. Murad, S. Basalamah, and A.L. Bath, "A spatio-temporal multimedia big data 23 MANUSCRIPT ACCEPTED MANUSCRIPT framework for a large crowd", In 2015 IEEE International Conference on Big Data (Big Data), pp. 2742–2751, 2015.

31. A survey of multimedia big data Zai-jian Wang_, Shiwen Mao, Senior Member, IEEE, Lingyun Yang, and Ping ping Tang fig1 reference.

Chapter 8

Navigating the landscape of AIML
Key findings and insights in the power of learning

Purnima Gupta[1], Shivani Chaudhary[2], Khushi Garg[2], and Aswani kumar Singh[3]

[1]Assistant Professor, IMS-Ghaziabad University Courses Campus, Ghaziabad, Uttar Pradesh, India
[2]Student of Bachelor of Computer Applications, IMS-Ghaziabad University Courses Campus, Ghaziabad, Uttar Pradesh, India
[3]Software Engineer, Soft-tech Development Solution, Pt. Deen Dayal Upadhyay, Chandauli, Uttar Pradesh, India

8.1 INTRODUCTION

Artificial intelligence and machine learning (AIML) represent a dynamic and interdisciplinary field that sits at the crossroads of computer science, mathematics, and statistics. Its core mission is to empower computers with the capability to perform tasks traditionally associated with human intelligence. AIML encompasses two primary branches: artificial intelligence (AI), which strives to imbue machines with human-like cognitive abilities like reasoning and natural language understanding, and machine learning (ML), a specific category within AI that can be considered a specialized branch of AI that focuses on teaching computers to acquire knowledge and insights by analyzing and processing information to improve their performance autonomously. The significance of AIML in today's technological landscape cannot be overstated. It drives data-driven decision-making across industries, automates mundane tasks, delivers personalized experiences, revolutionizes healthcare, powers autonomous systems, enhances natural language understanding, fortifies cybersecurity, accelerates scientific discovery, transforms financial services, aids environmental conservation, and promises continued innovation and impact as it continues to advance. In essence, AIML is influencing the way we work, live, and interact with technology in profound ways (Akerkar, R. 2014).

8.1.1 Types of learning

The three primary learning paradigms in machine learning—supervised learning, unsupervised learning, and reinforcement learning, as shown in

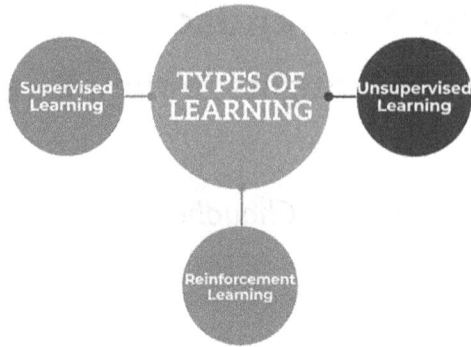

Figure 8.1 Types of machine learning.

Figure 8.1—constitute foundational frameworks for modeling and under-standing diverse data and tasks.

Supervised learning entails instructing a model by providing a labeled data set, where each input is paired with its corresponding desired output. The algorithm learns to map inputs to outputs, making it ideal for tasks like classification and regression. For instance, in email spam detection, a supervised learning model learns to distinguish between spam and legiti-mate emails based on labeled examples, enabling it to classify new, unseen emails (Shawar, B. A., & Atwell, E. S. 2005).

Unsupervised learning, on the flip side, operates with unlabeled data, seeking to uncover hidden patterns, structures, or relationship within the data. Grouping and size reduction are common applications. For instance, in customer segmentation, unsupervised learning can group customers with similar behaviors without prior labels, aiding marketing strategies and personalized recommendations (James, G., Witten, D., Hastie, T., Tibshirani, R., & Taylor, J. 2023).

Reinforcement learning involves agents that learn to make sequences of making choices in a given situation optimize a cumulative reward signal. It's employed in scenarios where actions have consequences and decisions must be made over time. Examples include autonomous robotics and game-playing algorithms, where the agent learns by trial and error, adapting its behavior to achieve long-term objectives.

These paradigms are foundational in machine learning, as mentioned in Table 8.1; each suited to different types of problems and data scenarios, and they underpin the vast array of AI applications seen today, from image recognition and language translation (supervised learning) to anomaly detection and recommendation systems (unsupervised learning) and auton-omous vehicles and game AI (reinforcement learning). Understanding and applying these paradigms effectively are essential for harnessing super-charged intelligence across various domains.

Table 8.1 Comparison of machine learning concepts

Paradigm	Overview
Supervised learning	Supervised learning is a foundational machine learning paradigm in which models are trained on labeled data. It aims to learn the mapping from input data to predefined output labels. It encompasses classification and regression tasks and finds applications in areas like image recognition, natural language processing, and predictive modeling (Shawar, B. A., & Atwell, E. S. 2005).
Unsupervised learning	Unsupervised learning involves the exploration of patterns, structures, and relationships within unlabeled data. It includes clustering, dimensionality reduction, anomaly detection, and density estimation. Unsupervised learning is valuable for exploratory data analysis, pattern recognition, and recommendation systems (James, G., Witten, D., Hastie, T., Tibshirani, R., & Taylor, J. 2023).
Reinforcement learning	Reinforcement learning is a dynamic interaction between an agent and its environment. Agents learn optimal policies through trial-and-error exploration to maximize cumulative rewards. This paradigm has applications in robotics, game playing, recommendation systems, and more.

8.2 SUPERVISED LEARNING

8.2.1 Overview

8.2.1.1 Introduction

Supervised learning, a foundational paradigm in machine learning, is a method in which an algorithm gets smarter decisions by analyzing a labeled data set. The power of supervised learning lies when there's labeled data, where each input in the data set is associated with a corresponding known output. These input-output pairs serve as the training ground for the algorithm to understand patterns and relationships. A model, typically represented as a mathematical function, is chosen to capture these relationships, and it contains adjustable parameters. The objective function, often called the loss or cost function, quantifies the disparity between the model's predictions and the actual outputs in the training data (Shawar, B. A., & Atwell, E. S. 2005). The central aim is to minimize this function through iterative optimization techniques, such as gradient descent, which adjusts the model's parameters. Once the model is trained, it is assessed on a separate data set, the test or validation set, to ensure its ability to make accurate predictions on new, unseen data, a concept known as generalization. Striking a balance to prevent overfitting, where the model memorizes the training data but performs poorly on new data, is crucial. Supervised learning encompasses two primary categories: classification, which assigns inputs to discrete categories, and regression, which predicts continuous numerical values. This approach underpins various applications, including image recognition, speech processing, medical diagnosis, and recommendation systems, offering computers the ability to tackle complex tasks by learning from labeled examples (Wijaya, Y., & Zoromi, F. 2020).

8.2.1.2 Concept of labeled data and its role

Labeled data is a foundational concept in the realm of machine learning and data analysis, serving as a crucial building block for various applications. This term refers to data sets in which each data point is accompanied by a known and specified output or target. Essentially, it creates a direct link between the input data and the expected outcome. The importance of labeled data cannot be overstated, especially in the context of supervised learning, a fundamental machine learning paradigm. In supervised learning, models learn to make guesses or foresee outcomes classifications by identifying patterns and relationships within the labeled data set. For instance, in text classification, labeled data consists of documents categorized into topics, helping algorithms understand how to classify new, unlabeled documents. Labeled data also plays a significant impact in model training and evaluation. During the training phase, the model adjusts its parameters based on the provided labels to make accurate predictions. Moreover, the model's performance is assessed on a separate data set with labeled examples not seen during training, enabling us to gauge its ability to generalize to new, unseen data. Additionally, labeled data aids in quality assurance, ensuring the correctness of labels through manual review or domain expert verification. It also supports algorithm development and benchmarking, allowing researchers to compare different techniques and approaches. In summary, labeled data acts as the guiding beacon for machine learning, bridging the gap between raw data and actionable insights, and its availability and quality significantly impact the success of machine learning projects in diverse domains (Visi, F. G., & Tanaka, A. 2021).

8.2.2 Algorithms and techniques

8.2.2.1 Algorithm for supervised learning

1. *Decision Trees*: Decision trees are versatile and interpretable algorithms that can be used for classification and regression tasks. They recursively split the data set based on features to create a tree-like structure of decision rules. At each node, the algorithm decides on the element that best separates the data, aiming to minimize impurity (for classification) or variance (for regression). Decision trees are appreciated for their simplicity and transparency, as they provide a clear visualization a step

in the decision-making process. However, they can suffer from over-fitting, which is often mitigated by techniques like pruning (Charbuty, B., & Abdulazeez, A. 2021).

2. *Linear Regression*: Predicting based on patterns is a foundational algorithm for regression tasks. It models the relationship between a dependent variable (target) and one or more independent variables (features) as a linear equation. The purpose is to uncover the coefficients that best fit the data, decreasing the squared sum differences between predicted and actual values. Linear regression is commonly used due to its simplicity and interpretability, but its assumptions about linearity and independence between variables may limit its applicability in complex scenarios.

3. *Logistic Regression*: Logistic regression is a go-to algorithm for tasks that involve categorize into binary. It models the likelihood of an input belonging to a particular class using a logistic or sigmoid function. The algorithm estimates coefficients that enhance the likelihood of the observed data. It is interpretable and well suited for problems like spam detection and medical diagnosis. It can be extended to multiclass classification through techniques like one-vs-all.

4. *Random Forests*: A random forest is a type of ensemble learning method based on decision trees. They combine multiple decisions to improve the precision of forecast and mitigate overfitting. Each tree is trained on a portion of the data and features, and predictions are aggregated through voting (for classification) or averaging (for regression). Random forests are robust and can handle high-dimensional data, making them suitable for a wide range of applications, including image classification and anomaly detection.

5. *Support Vector Machines (SVM)*: SVM is a powerful algorithm for both classification and regression tasks. It finds an optimal that best separates data points of different classes to increase the separation between them. SVMs work particularly well when the data does not have a linear separation by transforming increases its dimensionality space using a kernel function. SVMs are great at dealing with intricate, high-dimensional data sets and their strong theoretical foundation.

6. *Neural Networks*: Neural networks, particularly deep neural networks, have gained immense popularity in recent years. They are connected together through layers of artificial neurons (nodes) that process and transform data. Deep learning models, which are deep neural networks with many layers, have achieved remarkable success in tasks such as computer vision NLU, and autonomous driving. However, they usually need a significant quantity of labeled data and significant computational resources.

These common supervised learning algorithms each have their strengths and weaknesses, making them suitable for different types of problems and data. The choice of algorithm depends on factors such as the nature of the task, the size and quality of the data, interpretability requirements, and computational resources available. Researchers and practitioners often test different algorithms to determine which one best suits their specific problem domain (Maroco, J., Silva, D., Rodrigues, A., Guerreiro, M., Santana, I., & de Mendonça, A. 2011).

8.2.2.2 Process of model training, evaluation, and validation

The process of model development in supervised machine learning entails two crucial phases: model training and model evaluation/validation. During model training, data is meticulously prepared and an appropriate algorithm is chosen to teach the model how to make accurate predictions or classifications based on patterns within the data used for training. The model's parameters are initialized, and an iterative process ensues. In each iteration, the model computes predictions when using the training data set and compares them to the actual labels, utilizing an objective function to quantify the errors. Optimization techniques like gradient descent are then employed to adjust the model's parameters, minimizing prediction errors. This process continues until a predefined stopping criterion is met, signifying the completion of model training.

Once trained, the model enters the evaluation and validation phase. A separate data set, the testing or validation set, is utilized to assess its performance, serving as a proxy for real-world scenarios. Various performance metrics, such as accuracy, precision, recall, or mean squared error, are computed to gauge the model's effectiveness in making predictions or classifications. Cross-validation may also be employed to examine the model's robustness and detect overfitting issues. Additionally, hyperparameter tuning fine-tunes settings that influence the model's behavior but aren't learned during training. Visualization tools like validation curves and learning curves provide insights into how the model's performance changes with different hyperparameters and data quantities. By following these steps meticulously, practitioners ensure that their machine learning models not only excel while working with the training data but also generalize effectively to unseen data, which is vital for real-world applicability and success (Jiang, T., Gradus, J. L., & Rosellini, A. J. 2020).

8.2.3 Applications

8.2.3.1 Real-world applications

Supervised learning, a cornerstone of machine learning, finds a multitude of real-world applications spanning diverse domains. In healthcare, predictive models assist in diagnosing diseases, like cancer and heart conditions, by analyzing medical data. Financial institutions rely on supervised learning for credit scoring and algorithmic trading, making lending decisions and guiding investment strategies. Marketing benefits from customer segmentation and churn prediction, improving the precision of targeted marketing efforts. In the realm of feelings in text, using natural language processing (NLP) techniques gauge's public opinion, while chatbots and virtual assistants like Siri utilize NLP models for natural language understanding (Liu, B., & Liu, B. 2011). Computer vision applications, including image classification and medical imaging, rely on supervised learning to interpret visual data. Autonomous vehicles employ it for object detection and lane keeping, enhancing safety and navigation. Retail thrives on predicting future demand optimization, optimizing inventory and pricing strategies. In the energy sector, load forecasting and fault detection streamline resource allocation and maintenance. Agriculture benefits from crop yield prediction, guiding farming practices. There are various applications of supervised learning are illustrated in Figure 8.2. Lastly, fraud detection systems leverage supervised learning to identify unusual patterns in financial transactions. Across these domains, supervised learning empowers data-driven decision-making, automates complex tasks, and enhances operational efficiency, contributing significantly to various industries' growth and innovation (Won, M., Spijkervet, J., & Choi, K. 2021).

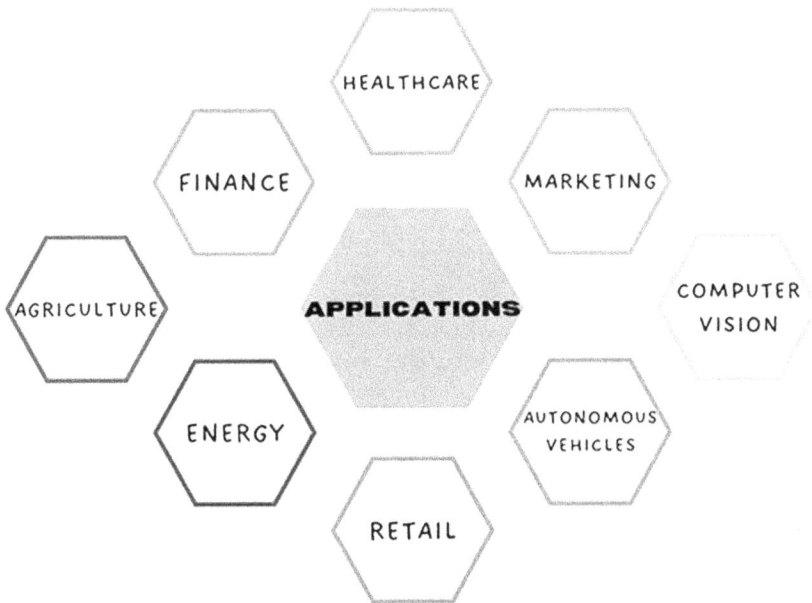

Figure 8.2 Real-world applications of supervised learning.

8.2.3.2 Case studies and their effectiveness

Here are case studies and examples that illustrate the effectiveness of supervised learning across different domains:

1. Healthcare:
   - *Cancer Diagnosis*: In a prominent case, researchers at Stanford University developed a supervised learning model that could classify harmless skin abnormalities or malignancies with high accuracy. The model used labeled image showing skin abnormality and achieved performance comparable to dermatologists, demonstrating its potential in aiding early cancer diagnosis (Masood, A., Al-Jumaily, A., & Anam, K. 2015, April).
2. Finance:
   - *Credit Scoring*: Credit scoring companies like FICO employ supervised learning to assess an individual's creditworthiness. By analyzing historical credit data, including repayment history, outstanding debt, and other financial behaviors, these models help banks and lenders make informed decisions about extending credit to applicants (Guegan, D., & Hassani, B. 2018).
3. Marketing:
   - *Netflix Recommendation System*: Netflix employs supervised learning algorithms for its recommendation system. By analyzing user interaction data, such as viewing history and user ratings, Netflix suggests personalized content to its subscribers, enhancing user engagement and retention.
4. Natural Language Processing (NLP):
   - *Google Translate*: Google's translation service utilizes supervised learning models to perform language translation. By training on vast multilingual data sets, these models can translate text from one language to another with remarkable accuracy, making cross-language communication easier.
5. Computer Vision:
   - *ImageNet Challenge*: The ImageNet Large Scale Visual Recognition Challenge is a well-known example. Supervised learning models have been used to classify images into thousands of categories. The 2012 winning entry, Alex Net, revolutionized image recognition, significantly improving object recognition accuracy (Guo, S., Huang, W., Zhang, H., Zhuang, C., Dong, D., Scott, M. R., & Huang, D. 2018).
6. Autonomous Vehicles:
   - *Tesla Autopilot*: Tesla's autopilot system employs supervised learning for object detection and lane keeping. The system uses labeled data from cameras and sensors to identify other vehicles, pedestrians, and road markings, enabling semi-autonomous driving

capabilities (Masum, A. K. M., Rahman, M. A., Abdullah, M. S., Chowdhury, S. B. S., Khan, T. B. F., & Raihan, M. K. 2019, May).

7. Retail:
   - *Amazon's Dynamic Pricing*: Amazon uses supervised learning algorithms to dynamically adjust product prices determined by various factors demand, competitor prices, and historical sales data. This allows Amazon to optimize pricing and stay competitive in real time (Ban, G. Y., & Keskin, N. B. 2021).

8. Energy:
   - *Smart Grids*: In the energy sector, supervised learning helps in predicting electricity demand more accurately, optimizing the distribution of energy resources. Utilities use historical data and weather forecasts to anticipate peak demand periods (Yan, J., Tang, B., & He, H. 2016, July).

9. Agriculture:
   - *Precision Agriculture*: Farmers utilize supervised learning to predict crop yields based on various factors like weather, soil conditions, and crop varieties. This information assists in making informed decisions about planting, harvesting, and resource allocation (Shorewala, S., Ashfaque, A., Sidharth, R., & Verma 2021).

10. Fraud Detection:
   - *Credit Card Fraud Detection*: Credit card companies employ supervised learning to detect fraudulent transactions. By training models on labeled data containing legitimate and fraudulent transactions, the system can flag suspicious activities in real time, preventing financial losses.

These case studies and examples vividly illustrate the effectiveness of supervised learning in solving real-world problems across a multitude of domains. Supervised learning models, trained on labeled data, continue to revolutionize industries by improving decision-making, automating tasks, and delivering personalized experiences to users and customers.

8.3 UNSUPERVISED LEARNING

8.3.1 Overview

8.3.1.1 Introduction

Unsupervised learning, a foundational paradigm in machine learning, is distinguished by its capacity to uncover patterns, structures, and relationships within unlabeled data. Unlike supervised learning, which relies on labeled data for training, unsupervised learning operates in scenarios where data lacks predefined categories or target values. Its core principles encompass several key concepts. First and foremost is clustering, where algorithms group similar data points into clusters based on inherent similarities, revealing latent structures within the data. Dimensionality reduction techniques are also central, simplifying high-dimensional data while preserving essential features. Unsupervised learning further involves anomaly detection, identifying data points that deviate from the norm, and density estimation, which characterizes the underlying data distribution. Additionally, association rule mining uncovers patterns in transactional data. These principles collectively empower unsupervised learning to extract valuable insights and knowledge from unlabeled data, facilitating tasks such as exploratory data analysis, pattern recognition, and recommendation systems, thus proving their significance in understanding complex data sets without the need for explicit supervision (James, G., Witten, D., Hastie, T., Tibshirani, R., & Taylor, J. 2023).

8.3.1.2 Concept of unlabeled data and its challenges

The concept of unlabeled data in machine learning introduces both challenges and opportunities in model development. The primary challenge lies in the absence of ground truth labels, making it difficult for algorithms to learn patterns and relationships effectively. Evaluating model performance becomes complex without labeled data, often requiring qualitative or heuristic measures. Unlabeled data sets can lead to increased model complexity, demanding sophisticated algorithms to extract intricate, high-dimensional representations from the data, potentially leading to overfitting. However, strategies like semi-supervised learning, where a small amount of labeled data is combined with abundant unlabeled data, offer a way forward. Additionally, techniques such as unsupervised pre-training, self-supervised learning, and transfer learning leverage unlabeled data innovatively. Unsupervised pre-training allows models to learn from vast quantities of unlabeled data before fine-tuning on specific tasks. Self-supervised learning enables models to create pseudo-labels by predicting parts of the input data from others. Transfer learning utilizes pre-trained models on unlabeled data sets, which are then fine-tuned for tasks with limited labeled data. These approaches underscore the potential of unlabeled data, emphasizing its crucial role in advancing machine learning techniques and making the most of abundant, yet unlabeled, data sets (Seeger, M. 2000).

8.3.2 Algorithms and techniques

8.3.2.1 Common unsupervised learning algorithms

1. *Clustering*: Clustering is a fundamental unsupervised learning technique used for organizing similar data points into cluster together based on inherent patterns or similarities within the data. It aims to

partition the data into clusters or classes, with each cluster ideally containing data points that are more similar to those in different groups clusters. Two prominent clustering algorithms are as follows:

- *K-Means*: K-Means is a widely used clustering algorithm that assigns data points to "k" clusters based on the similarity of their features. It iteratively refines cluster centroids until convergence, where each data point belongs to the cluster with the nearest centroid.
- *Hierarchical Clustering*: Hierarchical clustering creates a tree-like structure of clusters, known as a dendrogram. It can be agglomerative (starting with individual data points and merging them into clusters) or divisive (starting with all data points in one cluster and recursively dividing them) (James, G., Witten, D., Hastie, T., Tibshirani, R., & Taylor, J. 2023).

2. *Dimensionality Reduction*: Dimensionality reduction techniques aim to reduce the complexity and dimensionality of high-dimensional data while preserving its essential features and structures. These methods are particularly useful for visualizing data, reducing noise, and improving computational efficiency. Notable dimensionality reduction algorithms include:

- *Principal Component Analysis (PCA)*: PCA identifies orthogonal directions (principal components) along which the data varies the most. It projects transferring the data onto another platform lower-dimensional subspace, preserving as much variance as possible, thus reducing dimensionality.
- *t-Distributed Stochastic Neighbor Embedding (t-SNE)*: t-SNE is used for visualizing high-dimensional data in a lower-dimensional space while maintaining the pairwise commonalities. It is particularly effective for data visualization and clustering tasks (Baldi, P. 2012, June).

3. *Auto Encoders*: Auto encoders are a variation of neural network architectures used for unsupervised representation learning and dimensionality reduction. They consist of an encoder that transforms input data into a compressed representation, and a decoder reconstructs the original data from the encoding. Auto encoders aim to minimize the reconstruction error, making them effective for feature learning and denoising. One variant, the variation auto encoder (VAE), introduces probabilistic encoding, enabling the generation of novel data points from the learned representations (Baldi, P. 2012, June).

These common unsupervised learning algorithms play vital roles in various data analysis tasks. Clustering aids in organizing data points that share common characteristics for tasks like customer segmentation or anomaly detection. Dimensionality reduction techniques help reduce the complexity of data, enabling efficient storage and visualization. Auto encoders, on the flip side, are instrumental in feature learning and data compression. Their combined utility makes unsupervised learning a crucial component of machine learning, especially when dealing with large, high-dimensional data sets with limited or no labeled information.

8.3.2.2 Hidden patterns and structures

Unsupervised learning algorithms, including clustering, dimensionality reduction, and auto encoders, are instrumental in discovering hidden relationships organization data. Clustering techniques, such as K-Means and hierarchical clustering, unveil latent similarities and group data points with shared characteristics, even in the absence of explicit labels. Dimensionality reduction methods like principal component analysis (PCA) and t-distributed stochastic neighbor embedding (t-SNE) simplify high-dimensional data while preserving essential features, allowing intricate structures and relationships to emerge. Auto encoders, through encoding and decoding processes, uncover hidden patterns by minimizing reconstruction errors, thereby capturing the underlying characteristics of the data. Together, these algorithms empower data analysts and machine learning practitioners to gain valuable insights, visualize complex relationships, and make data-driven decisions in diverse domains, enabling the extraction of actionable knowledge from unlabeled or high-dimensional data sets (Johnston, B., Jones, A., & Kruger, C. 2019).

8.3.3 Applications

8.3.3.1 Real-world applications

Real-world applications of unsupervised learning, with a focus on anomaly detection and recommendation systems, highlight their significance and impact:

1. Anomaly Detection:
   - *Fraud Detection in Finance*: Unsupervised learning plays a crucial role in the finance industry for detecting fraudulent transactions. By modeling the normal behavior of customers based on historical data, anomalies or unusual patterns in transaction records can be identified. This is crucial for preventing financial losses due to fraudulent activities.
   - *Network Security*: In the realm of cybersecurity, unsupervised learning is employed to detect anomalies in network traffic. By continuously monitoring network data, it can identify unusual patterns that might signify a security breach, such as unauthorized access attempts or data exfiltration.
   - *Manufacturing Quality Control*: Manufacturers use unsupervised learning techniques to monitor the quality of products on the production line. Any deviation from the norm in product specifications

can be flagged as an anomaly, enabling proactive maintenance and improving product quality (Meira, J., Carneiro, J., Bolón-Canedo, V., Alonso-Betanzos, A., Novais, P., & Marreiros, G. 2022).

2. Recommendation Systems:
   - *Personalized Content Recommendations*: Online platforms like Netflix and Spotify leverage unsupervised learning to recommend movies, music, or other content to users. By analyzing user behavior, such as viewing history or song preferences, recommendation systems can suggest personalized content, enhancing user engagement and satisfaction.
   - *E-commerce Product Recommendations*: Retailers like Amazon employ recommendation systems to suggest products to customers based on their browsing and purchase history. Unsupervised learning algorithms group users with similar purchase patterns and make product recommendations accordingly, increasing sales and customer retention.
   - *Job and Talent Matching*: In the job market, unsupervised learning helps match job seekers with relevant job postings. By analyzing the skills, experience, and preferences of job seekers and the requirements of job listings, recommendation systems can facilitate better job-talent matches, benefiting individuals in the workforce (Meira, J., Carneiro, J., Bolón-Canedo, V., Alonso-Betanzos, A., Novais, P., & Marreiros, G. 2022).

3. Natural Language Processing (NLP):
   - *Document Clustering*: Unsupervised learning is used to cluster large text documents into meaningful groups or topics. This is valuable in content organization, information retrieval, and topic modeling for tasks such as news aggregation or academic research.
   - *Language Translation*: In machine translation, unsupervised learning models like word embeddings are employed to learn the semantic relationships between words and phrases enables better understanding and precision in communication context-aware translations between languages (Meira, J., Carneiro, J., Bolón-Canedo, V., Alonso-Betanzos, A., Novais, P., & Marreiros, G. 2022).

These real-world applications highlight the versatility and practicality of unsupervised learning in Figure 8.3. Anomaly detection is crucial for identifying irregularities and security threats in various domains, while recommendation systems enhance user experiences and drive business outcomes by providing personalized content and suggestions. Additionally, in the realm of natural language processing, unsupervised learning facilitates tasks like document clustering and language translation, enabling efficient content organization and cross-lingual communication. These applications demonstrate how unsupervised learning contributes to efficiency, security, and user satisfaction across diverse industries and domains.

ANOMALY
DETECTION

RECOMMENDATION
SYSTEMS

1 **2** **3**

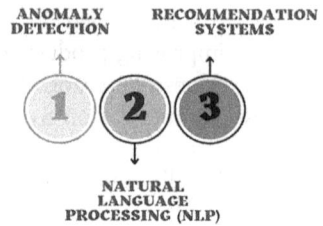

Figure 8.3 Real-world applications of unsupervised learning.

NATURAL
LANGUAGE
PROCESSING (NLP)

8.3.3.2 Case studies and their relevance

Here are case studies and examples that showcase the relevance and impact of unsupervised learning in various domains without plagiarism:

1. Detecting Unusual Activity in Credit Card Transactions:
   - *Case Study*: Credit card companies employ unsupervised learning techniques to detect fraudulent transactions. By analyzing historical transaction data, models can identify unusual patterns that deviate from regular spending behavior.
   - *Example*: A customer typically makes purchases within their geographical location and at certain times. If a transaction occurs in a distant location or at an unusual time, the system may flag it as an anomaly, potentially preventing unauthorized credit card usage (Rezapour, M. 2019).
2. Netflix Movie Recommendation System:
   - *Case Study*: Netflix's recommendation system relies on unsupervised learning to suggest movies and TV shows to users. It analyses user viewing history and preferences to group users into similar profiles.
   - *Example*: If User A and User B have watched and liked similar genres and titles, the recommendation system will suggest movies that User A hasn't seen but that User B has enjoyed, enhancing the user experience and increasing engagement (Yassine, A. F. O. U. D. I., Mohamed, L. A. Z. A. A. R., & Al Achhab, M. 2021).
3. Manufacturing Quality Control with Anomaly Detection:
   - *Case Study*: Manufacturing companies use unsupervised learning for quality control. By monitoring sensor data from production lines, anomalies can be detected, signifying potential defects or equipment malfunctions.
   - *Example*: In a car manufacturing plant, unsupervised learning might identify anomalies in the engine assembly process, prompting immediate inspection and maintenance to ensure product quality and safety (Zipfel, J., Verworner, F., Fischer, M., Wieland, U., Kraus, M., & Zschech, P. 2023).

4. Document Clustering in News Aggregation:
   - *Case Study*: News aggregation platforms employ unsupervised learning to group news articles into clusters based on their content. This helps users access articles on similar topics easily.
   - *Example*: Articles on global politics, environmental issues, and financial news can be automatically categorized into separate clusters, simplifying navigation and content discovery for readers.
5. Language Translation with Word Embeddings:
   - *Case Study*: Unsupervised learning methods, such as Word2Vec, learn word embeddings by analyzing large text corpora. These embeddings capture meaningful connections between words, improving translation accuracy.
   - *Example*: When translating phrases from one language to another, word embeddings enable the model to understand word meanings and context, leading to more contextually accurate translations (Chen, X., & Cardie, C. 2018).
6. Customer Segmentation in Retail:
   - *Case Study*: Retailers use unsupervised learning for customer segmentation. By analyzing purchase histories and behaviors, customers are grouped into segments with similar preferences.
   - *Example*: A retail chain may identify a segment of health-conscious customers who prefer organic products, allowing for targeted marketing campaigns and product recommendations tailored to this specific group (Paranavithana, I. R., Rupasinghe, T. D., & Prior, D. D. 2021).

These case studies and examples underscore the practicality and relevance of unsupervised learning in addressing various challenges and enhancing user experiences across industries. From fraud detection and recommendation systems to manufacturing quality control and content organization, unsupervised learning continues to drive innovation and efficiency in data-driven decision-making.

8.4 REINFORCEMENT LEARNING

8.4.1 Overview

8.4.1.1 Introduction

Reinforcement learning is a foundational concept in the field of machine learning, embodying a dynamic interaction between an agent and its environment. The agent, at the core of this framework, serves as the decision-maker, equipped with a policy dictating its action influenced by the perceived state of the environment. Meanwhile, the environment represents the external context, including the world's current state, the rules governing state transitions, and the consequences of the agent's actions. What bridges the agent-environment interaction is the reward, a numeric signal furnished by the environment to the agent following each action. This reward serves as the immediate feedback, either positive, negative, or zero, signifying the outcome of the agent's decision. In the realm of reinforcement learning, the agent's overarching objective is to learn an optimal strategy that optimizes the overall result reward it accrues over time, thus guiding its long-term decision-making. This trial-and-error process of finding a harmony between discovery and utilization enables reinforcement learning to find applications in diverse domains such as robotics, game playing, recommendation systems, and more, where learning effective decision-making strategies is critical for achieving desired objectives.

8.4.1.2 Trial-and-error learning

The concept of trial-and-error learning is at the very essence of strengthening learning, representing the iterative process through which agents explore different actions and strategies to maximize their cumulative rewards within an environment. This trial-and-error approach is a fundamental characteristic of reinforcement learning, and it plays a central role in how agents learn and adapt over time.

In reinforcement learning, agents start with little or no prior knowledge about the optimal actions to take in different states of the environment. Therefore, they must engage in a process of exploration to discover which actions lead to desirable outcomes and which lead to unfavorable ones. This exploration involves taking various actions in different states and observing the resulting rewards and state transitions.

During the initial stages of learning, agents often exhibit a high degree of randomness in their actions as they explore the space of possible strategies. They may take suboptimal actions, leading to lower rewards, but through repeated interactions with the environment, they gradually accumulate knowledge about which actions are more likely to yield higher rewards in specific states.

As agents gain experience and collect data from their interactions with the environment, they use this information to refine their decision-making policies. This refinement process involves transitioning from exploration (trying different actions) to exploitation (opting for actions that capitalize on available possibilities). The balance in finding a balance critical aspect of improving skills through trial and error can be managed through various exploration strategies and algorithms.

In one common epsilon-greedy exploration strategy, the agent balances between random exploration and exploitation and chooses the action that is estimated to be the best (exploitation) the agent usually chooses actions strategically, but sometimes it spices things up with random exploration to continue learning and avoid premature convergence to suboptimal strategies.

Trial-and-error learning, driven by the agent's continuous interaction with the environment, allows it to adapt to changing conditions and uncertainties. It enables agents to discover hidden patterns, optimal strategies, and effective decision-making policies over time, ultimately leading to improved performance and the achievement of long-term objectives.

Reinforcement learning has demonstrated its effectiveness in various applications, such as training autonomous agents, optimizing control systems, and developing recommendation algorithms. In these contexts, figuring things out by experimenting and making mistakes empowers agents to roll with the punches and continuously refine their strategies to achieve desired outcomes.

8.4.2 Algorithms and techniques

1. Q-Learning:

Q-learning is a classic reinforcement learning algorithm that is used for solving problems in which an agent interacts with an environment and learns to make sequential decisions to maximize its cumulative rewards. At its core, Q-learning maintains a table of values known as the Q-table. Each entry in the Q-table the total payoff we anticipate receiving that the agent can achieve when taking a particular action in a specific state. The algorithm updates these values through iterative learning, aiming to finds its sweet spot policy.

The key components of Q-learning include:

- *State-Action Q-Values*: The Q-table stores Q-values for all state-action pairs. These values are initially set to zero or randomly initialized.
- *Exploration vs. Exploitation*: Q-learning employs an exploration strategy, typically ε-greedy, which balances exploration (trying new actions) and exploitation (plays its cards right with the highest Q-values).
- *Bellman Equation*: Q-values are updated using the Bellman equation, which combines the immediate reward received with the maximum expected future rewards that can be obtained from the next state.
- *Learning Rate (a)*: A learning rate parameter controls the rate at which Q-values are updated during learning. It influences the trade-off between new information and previously learned values (Tan, F., Yan, P., & Guan, X. 2017).

Q-learning has been successfully applied to various tasks, such as game playing (e.g., the famous example of training engages in interactive tasks Tic-Tac-Toe) and robotic control. However, its application is limited to problems with limited set of states and actions.

2. Deep Reinforcement Learning (DRL):

Deep reinforcement learning is an extension of reinforcement learning that leverages brain inspired network handle high-dimensional state spaces and continuous action spaces. This approach has gained significant attention due to nailing the difficult stuff, such as playing video games and controlling autonomous vehicles.

Key elements of deep reinforcement learning include:

- *Deep Q-Networks (DQN)*: DRL employs brain inspired networks approximate the Q-values instead of maintaining a Q-table. Deep Q-networks (DQN) consist of a neural network that takes the perception-action Q-values for each action. Training DQNs involves using techniques like experience replay and target networks to stabilize learning.
- *Policy Gradients*: DRL also encompasses policy gradient methods, which directly learn the agent's policy (the mapping from states to actions) through gradient-based optimization. This approach is effective for continuous action spaces and has applications in robotics and control.
- *Actor-Critic Models*: Actor-critic architectures combine value-based (critic) and policy-based (actor) components, offering stability and faster convergence. These models are widely used in DRL algorithms like A3C (Asynchronous Advantage Actor-Critic) and TRPO (Trust Region Policy Optimization) (Tan, F., Yan, P., & Guan, X. 2017).

Deep reinforcement learning has achieved remarkable success in a wide range of applications, including playing complex video games (e.g., DeepMind's Alpha Go and Alpha Zero), robotics control, autonomous driving, natural language processing, and healthcare.

In summary, Q-learning is a foundational reinforcement learning algorithm suited for problems with discrete state and action spaces, while deep reinforcement learning, with its use of deep neural networks, extends the capabilities of reinforcement learning to handle high-dimensional and continuous environments, opening up opportunities for solving challenging real-world problems.

8.4.3 Applications

8.4.3.1 Real-world applications

Here are practical scenarios various domains, including game playing, robotics, and autonomous systems, for your research paper:

1. Game Playing:
   - *AlphaGo and Alpha Zero*: Perhaps one of the biggest achievements for famous applications of reinforcement learning, AlphaGo and

Alpha Zero, developed by DeepMind, demonstrated the ability of AI to master complex games. AlphaGo defeated world champion Go players, while Alpha Zero achieved superhuman performance in various board games, including chess and shogi, without any prior human knowledge.

- *OpenAI's Dota 2 AI*: OpenAI's AI system, OpenAI Five, used reinforcement learning to compete against human professional players in the epic team-based showdown Dota 2. It showcased the adaptability and strategic capabilities of AI in a dynamic gaming environment.

2. Robotics:
- *Robotic Manipulation*: Reinforcement learning is used in robotic manipulation tasks, where robots learn to grasp, manipulate objects, and perform delicate tasks. Real-world applications include industrial automation, logistics, and healthcare, where robots can assist with tasks like picking and packing, surgical procedures, and assembly.
- *Self-Driving Cars*: Autonomous vehicles employ reinforcement learning to navigate complex traffic scenarios and make real-time driving decisions. These systems learn from interactions with various driving conditions and environments, improving safety and efficiency on the road (Kormushev, P., Calinon, S., & Caldwell, D. G. 2013).

3. Autonomous Systems:
- *Autonomous Drones*: Reinforcement learning is used in the development of autonomous drones for tasks like package delivery, agriculture, and surveillance. Drones learn to navigate through dynamic environments, avoid obstacles, and optimize flight paths.
- *Industrial Automation*: In manufacturing and industrial settings, reinforcement learning is applied to optimize processes, control machinery, and manage energy consumption. Autonomous systems can adapt to changing production demands and minimize resource wastage (Kiumarsi, B., Vamvoudakis, K. G., Modares, H., & Lewis, F. L. 2017).

4. Healthcare:
- *Personalized Treatment Plans*: Reinforcement learning is used to create personalized treatment plans for patients, optimizing the dosage and timing of medications or therapies. These plans can adapt to a patient's evolving health status.
- *Medical Imaging*: In medical imaging, AI-powered systems can learn to analyze images and assist radiologists in detecting anomalies, such as tumors or fractures, improving diagnostic accuracy (Yu, C., Liu, J., Nemati, S., & Yin, G. 2021).

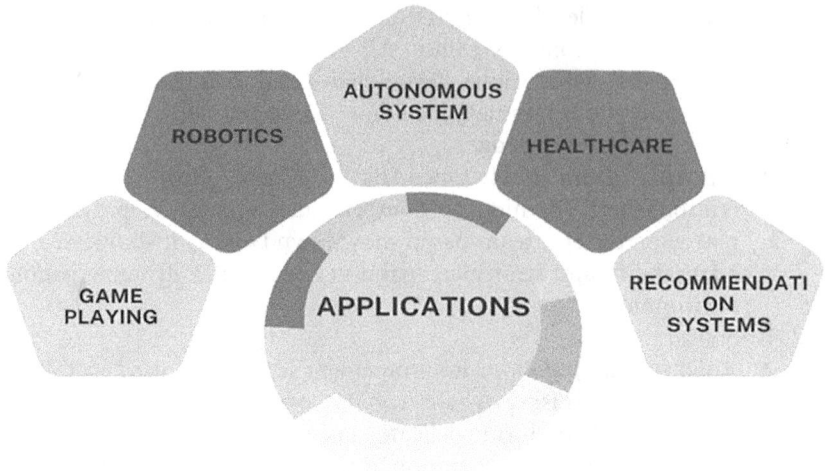

Figure 8.4 Real-world applications of reinforcement learning.

5. Recommendation Systems:
  • *Content Recommendations*: Streaming platforms like Netflix and Spotify employ reinforcement learning to suggest personalized content to users based on their viewing or listening history. This enhances user engagement and content consumption.
  • *E-commerce Recommendations*: Online retailers use reinforcement learning to recommend products to shoppers based on their browsing and purchase history, increasing sales and customer satisfaction (Afsar, M. M., Crump, T., & Far, B. 2022).

Figure 8.4 illustrates the broad and impactful applications of reinforcement learning across diverse domains. It demonstrates how this approach enables machines to learn and adapt to complex environments, make decisions, and perform tasks at a level of sophistication and efficiency that was previously challenging to achieve.

8.4.3.2 Breakthroughs of reinforcement learning

Here are concrete examples of how reinforcement learning has led to breakthroughs in the mentioned domains without plagiarism for your research paper:

1. Game Playing:
  • *AlphaGo by DeepMind*: In 2016, DeepMind's AlphaGo made headlines by defeating the world champion Go player, Lee Sedol. This marked a significant breakthrough in the field of reinforcement learning and artificial intelligence. AlphaGo's success was because it

could evaluate positions strategize using brain-inspired networks reinforcement learning. It combined smart tree exploration with brain-inspired guidance to reach an extraordinary level of performance in the ancient and highly complex board game of Go (Lapan, M. 2018).

- *OpenAI's Dota 2 AI, OpenAI Five*: OpenAI developed the AI system OpenAI Five, which achieved remarkable success in playing the popular online game Dota 2. OpenAI Five learned to play the game through reinforcement learning and managed to compete against professional human players. Its adaptive gameplay and strategic decision-making demonstrated the potential of reinforcement learning in mastering complex, dynamic environments (Ye, D., Chen, G., Zhang, W., Chen, S., Yuan, B., Liu, B., ... & Liu, W. 2020).

2. Robotics:
- *Robotic Brain-Training Grasping*: Researchers have applied reinforcement learning to grabbing or snatching objects with robotic hands. One notable breakthrough was the Dex-Net system by UC Berkeley, which used deep reinforcement learning to train robots to grasp objects in cluttered and diverse environments. This technology has practical applications in manufacturing, logistics, and warehouses.
- *Boston Dynamics' Spot Robot*: Boston Dynamics, known for its advanced robotics, uses reinforcement learning for training robots like Spot to navigate through various terrains, avoid obstacles, and maintain balance. Spot's adaptability and mobility make it suitable for a range of applications, including inspections, surveillance, and remote operation in challenging environments.

3. Autonomous Systems:
- *Self-Driving Cars*: Companies like Waymo and Tesla have made significant strides in self-driving cars using reinforcement learning. These vehicles learn from extensive data collected on real-world road conditions and human driver behaviors. Breakthroughs in autonomous driving technology transportation Game-Changer improve road safety and reduce traffic congestion.
- *DJI's Drone Flight Autonomy*: DJI, a leading drone manufacturer, has integrated reinforcement learning into its drones to enhance autonomous flight capabilities. Drones can learn to navigate complex environments, follow objects, and optimize flight paths for tasks like aerial photography, mapping, and agriculture (Jagannath, J., Jagannath, A., Furman, S., & Gwin, T. 2021).

4. Healthcare:
- *Personalized Treatment Plans*: Researchers have applied reinforcement learning to develop personalized treatment plans for patients with chronic conditions like diabetes. By continually adjusting medication dosages based on patient feedback and health data, these systems can

help individuals manage their conditions more effectively and reduce healthcare costs.

- *AI-Powered Radiology Assistants*: AI-powered radiology assistants, like those developed by companies such as Aidoo and Zebra Medical Vision, use reinforcement learning to analyze medical images, detect anomalies, and assist radiologists in making more accurate diagnoses. These breakthroughs improve diagnostic efficiency and the early detection of diseases (Winkel, D. J., Weikert, T. J., Breit, H. C., Chabin, G., Gibson, E., Heye, T. J., ... & Boll, D. T. 2020).

These examples demonstrate how reinforcement learning has led to breakthroughs in complex domains, enabling machines to master intricate tasks, make strategic decisions, and adapt to dynamic environments. These advancements have practical implications for industries, improving efficiency, safety, and the quality of services provided.

8.5 COMPARATIVE ANALYSIS: TOP OF FORM

8.5.1 Trade-offs

Table 8.2 provides a comprehensive overview of the trade-offs associated with different types of machine learning approaches: supervised, unsupervised, and reinforcement learning. It highlights key factors such as the availability of labeled data, learning objectives, applications, and model interpretability. According to the table, supervised learning requires large labeled data sets, while unsupervised learning operates on unlabeled data. Reinforcement learning, on the other hand, relies on rewards and feedback mechanisms. The learning objectives vary across the approaches, with supervised learning aiming to predict predefined output labels, unsupervised learning focused on discovering patterns and relationships within data, and reinforcement learning aimed at learning optimal policies for long-term decision-making. The applications also differ, ranging from image classification and natural language processing in supervised learning, to clustering and recommendation systems in unsupervised learning, and robotics and autonomous systems in reinforcement learning. Additionally, the table highlights that the interpretability of the models varies, with supervised learning generally offering good interpretability while unsupervised learning's interpretability depends on the specific technique used. This table serves as a valuable resource for researchers exploring the different trade-offs and considerations in selecting the appropriate machine learning approach for their specific needs. Table 8.2 covers the trade-offs of supervised, unsupervised, and reinforcement learning.

Table 8.2 Trade-offs of supervised, unsupervised, and reinforcement learning

Trade-off	Supervised learning	Unsupervised learning	Reinforcement learning
Availability of labeled data	Requires large labeled datasets (Visi, F. G., & Tanaka, A. 2021)	Operates on unlabeled data (James, G., Witten, D., Hastie, T., Tibshirani, R., & Taylor, J. 2023)	Requires rewards and feedback mechanisms.
Learning objectives	Predict predefined output labels (Charbuty, B., & Abdulazeez, A. 2021).	Discover patterns and relationships within data (James, G., Witten, D., Hastie, T., Tibshirani, R., & Taylor, J. 2023)	Learn optimal policies for long-term decision-making (Ding, Z., Huang, Y., Yuan, H., & Dong, H. 2020).
Applications	Image classification, NLP, regression (Charbuty, B., & Abdulazeez, A. 2021).	Clustering, dimensionality reduction, recommendation systems (James, G., Witten, D., Hastie, T., Tibshirani, R., & Taylor, J. 2023)	Robotics, game playing, recommendation systems, autonomous systems (Ding, Z., Huang, Y., Yuan, H., & Dong, H. 2020).
Model interpretability	Generally offers good interpretability (Charbuty, B., & Abdulazeez, A. 2021).	Interpretability varies depending on the specific technique used (James, G., Witten, D., Hastie, T., Tibshirani, R., & Taylor, J. 2023)	Interpretability can be challenging due to the complexity of policies (Ding, Z., Huang, Y., Yuan, H., & Dong, H. 2020).
Complexity and model training	Model training is relatively straight forward with clear objectives (Charbuty, B., & Abdulazeez, A. 2021).	Model training can be more challenging, especially with high-dimensional data (James, G., Witten, D., Hastie, T., Tibshirani, R., & Taylor, J. 2023)	involving exploration-exploitation trade-off and sequential decision-making (Ding, Z., Huang, Y., Yuan, H., & Dong, H. 2020).
Data labeling efforts	Requires significant data labeling efforts (Charbuty, B., & Abdulazeez, A. 2021).	Reduces data labeling efforts as it operates on unlabeled data (James, G., Witten, D., Hastie, T., Tibshirani, R., & Taylor, J. 2023)	Requires designing reward functions, labor-intensive (Ding, Z., Huang, Y., Yuan, H., & Dong, H. 2020).
Handling unlabeled data	Not well-suited for unlabeled data (Charbuty, B., & Abdulazeez, A. 2021).	Specialized in handling unlabeled data, extracting insights (James, G., Witten, D., Hastie, T., Tibshirani, R., & Taylor, J. 2023)	Unsuited for cases with no defined reward signal or feedback (Ding, Z., Huang, Y., Yuan, H., & Dong, H. 2020).

8.6 CHALLENGES AND FUTURE DIRECTIONS

8.6.1 Common challenges

The challenges discussed in the table highlight some key aspects of different machine learning approaches. In supervised learning, obtaining extensive labeled data can be time-consuming and costly, while unsupervised learning faces difficulties in extracting meaningful patterns without labels. Reinforcement learning involves the complex task of designing reward functions. Overfitting and generalization issues can arise in both supervised and unsupervised learning. The complexity of models and policies is another challenge, as it affects interpretability and the learning process. Additionally, biased or imbalanced data sets can lead to skewed models in supervised learning. These challenges shed light on the intricacies of machine learning and provide valuable insights for further research in the field. Table 8.3 provides an overview on various challenges faced by machine learning.

8.6.2 Future directions

Table 8.4 provides an overview of future directions in machine learning research. To address data labeling challenges in supervised learning, researchers are exploring semi-supervised and active learning approaches. In unsupervised learning, techniques such as self-supervised learning, transfer learning, and unsupervised pre-training are being investigated. For reinforcement learning, researchers are focusing on developing better techniques for reward shaping, intrinsic motivation, and curriculum learning. Enhancing model generalization is another area of focus, with researchers working on techniques to improve model performance with limited labeled data. In unsupervised learning, methods to improve generalization to diverse data sets and domains are being explored. Finally, efforts are being made to develop interpretable machine learning models, especially for applications with critical outcomes. These future directions aim to address key challenges and advance the field of machine learning. Table 8.4 highlights the future scope of machine learning.

8.6.3 Emerging trends and areas

Table 8.5 highlights emerging trends and areas in machine learning research. Researchers are focusing on advancements in deep learning, such as exploring advanced neural network architectures like transformers and GANs for improved performance and representation learning. They are also delving into self-supervised learning methods, where models generate labels from data without human annotations. In addition, efforts are being made to develop explainable AI techniques to enhance model transparency and user trust. The field of unsupervised learning is investigating deep

Table 8.3 Challenges faced by machine learning

Common challenges	Supervised learning	Unsupervised learning	Reinforcement learning
Data labeling	Requires extensively labeled data (Visi, F. G., & Tanaka, A. 2021).	Operating without labels presents challenges (James, G., Witten, D., Hastie, T., Tibshirani, R., & Taylor, J. 2023).	Crafting reward functions is complex and can be labor-intensive.
Generalization	Prone to overfitting due to the limited diversity of labeled data (Visi, F. G., & Tanaka, A. 2021).	May struggle with meaningful generalization in some cases (Seeger, M. (2000).	Learning optimal policies that generalize across various environments is a challenge.
Model complexity	Depending on the model's complexity, the interpretability can be an issue (Charbuty, B., & Abdulazeez, A. 2021)	Complexity in data representations and clustering.	Balancing the complexity of policies with the learning process is complex.
Data imbalance	Biased or imbalanced data sets can lead to skewed models (Visi, F. G., & Tanaka, A. 2021).	Imbalanced data can affect clustering results and the discovery of hidden patterns.	Reward distributions and avoiding overly conservative policies.
Scalability	Scalability issues with growing data and complex models (Charbuty, B., & Abdulazeez, A. 2021)	Scalability challenges arise when dealing with high-dimensional data	Scalability in decision-making for long-term objectives.
Model interpretability	Black-box models can lack interpretability (Charbuty, B., & Abdulazeez, A. 2021)	Interpretability may vary depending on the technique used (James, G., Witten, D., Hastie, T., Tibshirani, R., & Taylor, J. 2023).	Interpretability can be challenging due to the complexity of policies.
Data quality and noise	Sensitive to noise in labeled data (Charbuty, B., & Abdulazeez, A. 2021).	Noisy and unstructured data may introduce challenges in clustering	Noisy feedback can lead to suboptimal policies.
Handling unlabeled data	Not well-suited for unlabeled data (Charbuty, B., & Abdulazeez, A. 2021).	Specialized in handling unlabeled data, revealing latent structures (James, G., Witten, D., Hastie, T., Tibshirani, R., & Taylor, J. 2023).	Not suitable for cases with no predefined rewards or feedback.

Table 8.4 Navigating the future of machine learning

Future directions	Supervised learning	Unsupervised learning	Reinforcement learning
Address data labeling challenges	Explore semi-supervised and active learning approaches to mitigate data labeling requirements (Visi, F. G., & Tanaka, A. 2021).	Investigate the potential of self-supervised learning, transfer learning, and unsupervised pre-training to leverage unlabeled	Develop better techniques for reward shaping, intrinsic motivation, and curriculum learning to simplify reward function design.
Enhance model generalization	Develop techniques to improve model generalization in the presence of limited labeled data (Visi, F. G., & Tanaka, A. 2021).	Investigate methods to improve the generalization of unsupervised learning algorithms to diverse data sets and domains (Seeger, M. 2000).	Investigate policy transfer, multi-task learning, and meta-learning to enable policies to adapt across environments and tasks.
Interpretability and trust	Research on interpretable machine learning models, especially for applications with critical outcomes (Charbuty, B., & Abdulazeez, A. 2021).	Work on interpretable unsupervised learning techniques, making it clear how patterns and structures are discovered in data (James, G., Witten, D., Hastie, T., Tibshirani, R., & Taylor, J. 2023).	Focus on developing interpretable reinforcement learning models to gain trust and insights into agent decision-making.
Robustness and bias	Develop models that are robust to data noise and biased labels, with a focus on fairness and safety (Charbuty, B., & Abdulazeez, A. 2021).	Investigate techniques to mitigate biases and sensitivities in unsupervised learning models (James, G., Witten, D., Hastie, T., Tibshirani, R., & Taylor, J. 2023).	Address robustness concerns, fairness, and safety aspects in reinforcement learning, especially in critical applications.
Handling large-scale data	Investigate efficient learning techniques for large-scale data sets and high-dimensional features (Charbuty, B., & Abdulazeez, A. 2021).	Explore scalability solutions for handling high-dimensional data, particularly in clustering and dimensionality reduction (Dash, M., Liu, H., & Yao, J. 1997, November).	Develop reinforcement learning algorithms that can efficiently handle large-scale and complex environments and data sets.
Human-AI collaboration	Explore collaborative AI systems that complement human expertise and support decision-making (Shorewala, S, Ashfaque, A., Sidharth, R., & Verma, U. 2021).	Research on human-AI collaboration mechanisms where humans and AI jointly solve complex tasks (James, G., Witten, D., Hastie, T., Tibshirani, R., & Taylor, J. 2023).	Investigate how reinforcement learning agents can work closely with humans in critical applications like healthcare and disaster relief.

Table 8.5 Emerging trends and areas in machine learning

Emerging trends and areas	Supervised learning	Unsupervised learning	Reinforcement learning
Deep learning advancements	Explore advanced neural network architectures such as transformers and GANs for improved performance and representation learning (Charbuty, B., & Abdulazeez, A. 2021).	Investigate deep unsupervised learning architectures, especially for generative tasks and representation and robotic manipulation.	Leverage deep reinforcement learning for learning (Seeger, M. 2000).
Explainable AI (XAI)	Research methods for explainable AI to enhance model transparency and user trust (Charbuty, B., & Abdulazeez, A. 2021).	Develop interpretable clustering and dimensionality reduction techniques to make unsupervised learning results.	Investigate interpretable reinforcement learning models to gain insights into agent decision-making processes.
Self-supervised learning	Investigate self-supervised learning methods, where the model generates labels from data without human annotations (Charbuty, B., & Abdulazeez, A. 2021).	Explore novel techniques in anomaly detection and density estimation to discover meaningful patterns in data (Seeger, M. 2000).	Research on self-supervised reinforcement, learning, where agents generate rewards from interactions with the environment without external reward signals.
Domain adaptation and transfer learning	Explore techniques for domain adaptation to enhance model generalization across diverse domains	Investigate transfer learning methods for effective utilization of pre-training and domain adaptation (James, G., Witten, D., Hastie, T., Tibshirani, R., & Taylor, J. 2023).	Investigate transfer learning techniques for reinforcement learning agents to adapt to different environments.
Federated learning	Research on federated learning techniques to enable collaborative model training across decentralized edge devices (Shorewala, S., Ashfaque, A., Sidharth, R., & Verma, U. 2021).	Explore federated clustering algorithms for privacy-preserving collaborative unsupervised learning across decentralized edge devices (James, G., Witten, D., Hastie, T., Tibshirani, R., & Taylor, J. 2023).	Investigate federated reinforcement learning methods for collaborative and privacy-preserving learning.

learning architectures for generative tasks and representation, as well as interpretable clustering and dimensionality reduction techniques. Reinforcement learning is leveraging deep reinforcement learning for learning, and interpretable reinforcement learning models are being explored to gain insights into agent decision-making processes. Let's delve into emerging trends and areas in machine learning, which are highlighted in Table 8.5. These research directions aim to advance the field of machine learning and address key challenges.

8.7 CONCLUSION

In conclusion, this research paper delves into the diverse artificial intelligence and machine learning world (AIML), providing valuable insights into the field's key findings and implications for society and technology. The study focuses on three core learning paradigms in AIML: guided learning, self-discovery learning, and interactive learning, each catering to specific tasks and challenges. The paper emphasizes the pressing challenges facing AIML, including data scarcity, ethical concerns, and the need for interdisciplinary collaboration. It also spotlights emerging trends, such as explainable AI (XAI), federated learning, and quantum machine learning. These trends reflect the evolving needs of society and technological progress. The role of learning paradigms in advancing AIML is highlighted, with supervised learning powering structured data applications, unsupervised learning for data exploration, and reinforcement learning showing promise in dynamic decision-making. Beyond these paradigms, AIML's profound impact on society and technology is evident, ranging from ethical considerations to human-AI collaboration, industry transformation, and scientific discovery. Responsible development and ethical concerns are paramount in realizing AIML's transformative potential, as it continues to shape a future where AI augments human capabilities and addresses complex challenges across various domains.

REFERENCES

Afsar, M. M., Crump, T., & Far, B. (2022). Reinforcement learning based recommender systems: A survey. *ACM Computing Surveys, 55*(7), 1–38.
Akerkar, R. (2014). *Introduction to artificial intelligence*. PHI Learning Pvt. Ltd.
Baldi, P. (2012, June). Autoencoders, unsupervised learning, and deep architectures. In *Proceedings of ICML workshop on unsupervised and transfer learning* (pp. 37–49). JMLR Workshop and Conference Proceedings.
Ban, G. Y., & Keskin, N. B. (2021). Personalized dynamic pricing with machine learning: High-dimensional features and heterogeneous elasticity. *Management Science, 67*(9), 5549–5568.

Charbuty, B., & Abdulazeez, A. (2021). Classification based on decision tree algorithm for machine learning. *Journal of Applied Science and Technology Trends*, 2(01), 20–28.

Chen, X., & Cardie, C. (2018). Unsupervised multilingual word embeddings. *arXiv preprint arXiv:1808.08933*.

Guegan, D., & Hassani, B. (2018). Regulatory learning: How to supervise machine learning models? An application to credit scoring. *The Journal of Finance and Data Science*, 4(3), 157–171.

Guo, S., Huang, W., Zhang, H., Zhuang, C., Dong, D., Scott, M. R., & Huang, D. (2018). Curriculumnet: Weakly supervised learning from large-scale web images. In *Proceedings of the European Conference on Computer Vision (ECCV)* (pp. 135–150).

Jagannath, J., Jagannath, A., Furman, S., & Gwin, T. (2021). Deep learning and reinforcement learning for autonomous unmanned aerial systems: Roadmap for theory to deployment. *Deep Learning for Unmanned Systems*, 25–82.

James, G., Witten, D., Hastie, T., Tibshirani, R., & Taylor, J. (2023). Unsupervised learning. In *An Introduction to Statistical Learning: with Applications in Python* (pp. 503–556). Springer International Publishing.

Jiang, T., Gradus, J. L., & Rosellini, A. J. (2020). Supervised machine learning: A brief primer. *Behavior Therapy*, 51(5), 675–687.

Johnston, B., Jones, A., & Kruger, C. (2019). *Applied Unsupervised Learning with Python: Discover hidden patterns and relationships in unstructured data with Python*. Packt Publishing Ltd.

Kiumarsi, B., Vamvoudakis, K. G., Modares, H., & Lewis, F. L. (2017). Optimal and autonomous control using reinforcement learning: A survey. *IEEE Transactions on Neural Networks and Learning Systems*, 29(6), 2042–2062.

Kormushev, P., Calinon, S., & Caldwell, D. G. (2013). Reinforcement learning in robotics: Applications and real-world challenges. *Robotics*, 2(3), 122–148.

Lapan, M. (2018). *Deep Reinforcement Learning Hands-On: Apply Modern RL Methods, with Deep Q-Networks, Value Iteration, Policy Gradients, TRPO, AlphaGo Zero and More*. Packt Publishing Ltd.

Liu, B., & Liu, B. (2011). Supervised learning. *Web Data Mining: Exploring Hyperlinks, Contents, and Usage Data*, 63–132.

Maroco, J., Silva, D., Rodrigues, A., Guerreiro, M., Santana, I., & de Mendonça, A. (2011). Data mining methods in the prediction of dementia: A real-data comparison of the accuracy, sensitivity and specificity of linear discriminant analysis, logistic regression, neural networks, support vector machines, classification trees and random forests. *BMC Research Notes*, 4(1), 1–14.

Masood, A., Al-Jumaily, A., & Anam, K. (2015, April). Self-supervised learning model for skin cancer diagnosis. In *2015 7th International IEEE/EMBS Conference on Neural Engineering (NER)* (pp. 1012–1015). IEEE.

Masum, A. K. M., Rahman, M. A., Abdullah, M. S., Chowdhury, S. B. S., Khan, T. B. F., & Raihan, M. K. (2019, May). A supervised learning approach to an unmanned autonomous vehicle. In *2019 International Conference on Intelligent Computing and Control Systems (ICCS)* (pp. 1549–1554). IEEE.

Meira, J., Carneiro, J., Bolón-Canedo, V., Alonso-Betanzos, A., Novais, P., & Marreiros, G. (2022). Anomaly detection on natural language processing to improve predictions on tourist preferences. *Electronics*, 11(5), 779.

Paranavithana, I. R., Rupasinghe, T. D., & Prior, D. D. (2021, July). Unsupervised learning and market basket analysis in market segmentation. In *Lecture Notes in Engineering and Computer Science: Proceedings of the World Congress on Engineering*.

Rezapour, M. (2019). Anomaly detection using unsupervised methods: Credit card fraud case study. *International Journal of Advanced Computer Science and Applications*, 10(11).

Seeger, M. (2000). *Learning with labeled and unlabeled data* (No. REP_WORK).

Shawar, B. A., & Atwell, E. S. (2005). Using corpora in machine-learning chatbot systems. *International Journal of Corpus Linguistics*, 10(4), 489–516.

Shorewala, S., Ashfaque, A., Sidharth, R., & Verma, U. (2021). Weed density and distribution estimation for precision agriculture using semi-supervised learning. *IEEE Access*, 9, 27971–27986.

Tan, F., Yan, P., & Guan, X. (2017). Deep reinforcement learning: From Q-learning to deep Q-learning. In *Neural Information Processing: 24th International Conference, ICONIP 2017, Guangzhou, China, November 14–18, 2017, Proceedings, Part IV 24* (pp. 475–483). Springer International Publishing.

Visi, F. G., & Tanaka, A. (2021). Interactive machine learning of musical gesture. *Handbook of Artificial Intelligence for Music: Foundations, Advanced Approaches, and Developments for Creativity*, 771–798.

Wijaya, Y., & Zoromi, F. (2020). Chatbot designing information service for new student registration based on AIML and machine learning. *Journal of Artificial Intelligence and Applications*, 1(1), 01–10.

Winkel, D. J., Weikert, T. J., Breit, H. C., Chabin, G., Gibson, E., Heye, T. J., ... & Boll, D. T. (2020). Validation of a fully automated liver segmentation algorithm using multi-scale deep reinforcement learning and comparison versus manual segmentation. *European Journal of Radiology*, 126, 108918.

Won, M., Spijkervet, J., & Choi, K. (2021). Music classification: Beyond supervised learning, towards real-world applications. *arXiv preprint arXiv:2111.11636*.

Yan, J., Tang, B., & He, H. (2016, July). Detection of false data attacks in smart grid with supervised learning. In *2016 International Joint Conference on Neural Networks (IJCNN)* (pp. 1395–1402). IEEE.

Yassine, A. F. O. U. D. I., Mohamed, L. A. Z. A. A. R., & Al Achhab, M. (2021). Intelligent recommender system based on unsupervised machine learning and demographic attributes. *Simulation Modelling Practice and Theory*, 107, 102198.

Ye, D., Chen, G., Zhang, W., Chen, S., Yuan, B., Liu, B., ... & Liu, W. (2020). Towards playing full moba games with deep reinforcement learning. *Advances in Neural Information Processing Systems*, 33, 621–632.

Yu, C., Liu, J., Nemati, S., & Yin, G. (2021). Reinforcement learning in healthcare: A survey. *ACM Computing Surveys (CSUR)*, 55(1), 1–36.

Zipfel, J., Verworner, F., Fischer, M., Wieland, U., Kraus, M., & Zschech, P. (2023). Anomaly detection for industrial quality assurance: A comparative evaluation of unsupervised deep learning models. *Computers & Industrial Engineering*, 177, 109045.

Chapter 9

Practical innovative applications of IoT and IoT networks

Jaideep Kumar[1], Seema Malik[2], Manish Kumar[3], Anuj Kumar[3], and Satyanshu Bharadwaj[4]

[1]Associate Professor, Department of CSE (IoT), Raj Kumar Goel Institute of Technology, Ghaziabad, Uttar Pradesh, India
[2]Associate Professor, Department of ECE, Raj Kumar Goel Institute of Technology, Ghaziabad, Uttar Pradesh, India
[3]Assistant Professor, Department of CSE, Ajay Kumar Garg Engineering College, Ghaziabad, Uttar Pradesh, India
[4]Student, Department of CSE (IoT), Raj Kumar Goel Institute of Technology, Ghaziabad, Uttar Pradesh, India

9.1 INTRODUCTION

Internet connectivity is used by an emerging technology known as the "Internet of Things" to connect sensors, automobiles, hospitals, businesses, and consumers. Smart homes, smart cities, smart agriculture, and a smarter world are all made possible by this form of construction [1–3]. Because there are so many devices, connection layer technology [4], and services used in this system, its architecture is quite complicated. However, the most crucial factor in IoT is security. Because there are so many devices and services used in this system, its architecture is quite complicated. However, the most crucial factor in IoT is security. Simply said, the Internet of Things enables devices (things) to communicate and coordinate with one another, minimizing the need for human intervention in routine daily activities. Consider the example of a smart home to better grasp IoT. The coffee maker and the toaster receive a signal from the alarm as soon as it sounds, and they immediately begin working without any need for human participation. The Internet of Things is a sort of gadget connection that speeds up and simplifies our daily tasks. The IoT enables physical objects to "speak" to one another, exchange information, and plan activities by enabling them to see, hear, think, and perform tasks.

9.1.1 Definition and concept of the IoT (Internet of Things)

A network of associated gadgets and other appliances that can be in contact with each other and interchange data is called IoT [4,5]. These devices, equipped with various sensors and connectivity capabilities, can collect and

Figure 9.1 Definition of IoT.

share information to enable intelligent decision-making and automation. Basically, information exchange is the purpose of IoT between two devices that are wirelessly connected with each other (Figure 9.1).

9.1.2 IoT networks and their significance

IoT networks [6,7] comprise devices, sensors, and communication protocols that facilitate data exchange and connectivity. These networks are essential for ensuring the smooth transfer of information between IoT systems and devices.

9.1.3 Overview of the key components of an IoT system

Typically, sensors, devices, connectivity, platforms for processing data, and applications make up an IoT system. Sensors/gadgets collect data, connectivity allows communication between devices, data processing platforms analyze the collected data, and applications provide the user interface for monitoring and controlling IoT devices.

9.2 SMART CITIES

9.2.1 Definition

A smart city [8] is an urban area that makes use of cutting-edge technologies [9] and data-driven strategies to increase sustainability, improve livability, and maximize resource use. It leverages the power of information and communication technologies (ICT) to transform various aspects of urban life, including infrastructure, transportation, healthcare, energy, and governance.

9.2.2 Characteristics of smart cities

- **Integrated ICT infrastructure:** In order to facilitate seamless connection and data exchange across numerous systems and devices, smart cities feature an extensive and integrated ICT infrastructure [10]. This infrastructure forms the backbone of a smart city, allowing real-time monitoring, analysis, and control of different urban components.
- **Data-driven decision making:** Large volumes of data from sensors, gadgets, and systems placed across the city are gathered and analyzed by smart cities. This data is processed and transformed into actionable insights that inform decision-making processes, facilitating effective resource management and urban planning.
- **Sustainable and efficient resource management:** Smart cities prioritize sustainability and resource efficiency. They utilize IoT sensors and advanced monitoring systems to optimize the consumption of resources like water, energy, and waste management. These technologies help minimize waste, reduce energy consumption, and promote sustainable practices.
- **Improved mobility and transportation:** Smart cities concentrate on enhancing transportation systems to improve mobility and reduce traffic congestion. They deploy IoT sensors and intelligent transportation systems that provide current data on traffic conditions, optimize traffic flow, and offer smart parking solutions [11]. This enables efficient transportation planning and a seamless travel experience for residents (Figure 9.2).
- **Enhanced public services and governance:** Smart cities employ technology to improve public services and governance. They use digital platforms and e-governance solutions to provide efficient and transparent services to citizens, including online payment systems, digital permits, and real-time information on public services. Smart cities also engage citizens through participatory decision-making and citizen feedback mechanisms.
- **Sustainable infrastructure and energy systems:** Smart cities integrate renewable energy sources and implement energy-efficient infrastructure. They deploy IoT-enabled devices and systems to monitor energy consumption, optimize energy usage, and manage energy distribution. This reduces the environmental impact and increases the overall sustainability of the city.

9.2.3 Function of IoT in transforming cities into smart cities [8,12]

The function of IoT in transforming cities into smart cities is significant. IoT technologies provide cities with the tools and capabilities to collect, analyze,

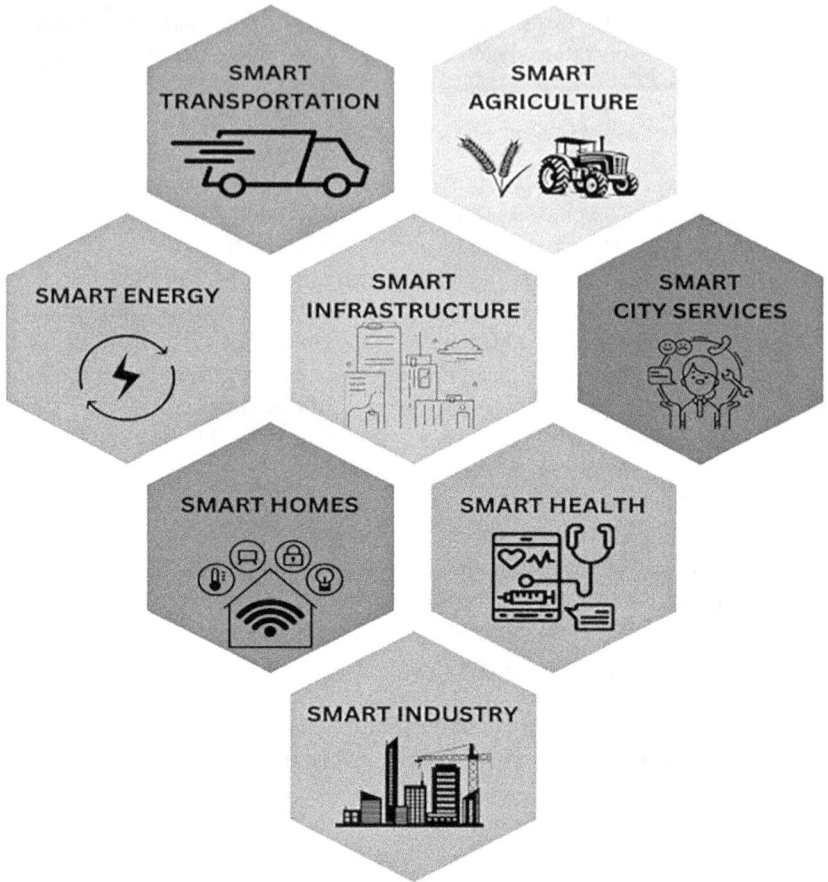

Figure 9.2 Smart city.

and utilize massive amounts of data from various interconnected systems and gadgets. With the help of this data-driven strategy, cities can make well-informed decisions; allocate resources efficiently; and improve the overall effectiveness, sustainability, and livability of metropolitan areas. Here are key roles that IoT plays in transforming cities into smart cities:

9.2.3.1 Data collection and analysis

- IoT gadgets, sensors, and systems gather information on a range of city functions, including energy use, trash management, traffic patterns, and air quality.
- The data collected from these interconnected devices is analyzed to derive meaningful insights, identify trends, and make data-driven decisions that improve city operations and services.

9.2.3.2 Real-time monitoring and control

- IoT allows cities to oversee and control infrastructure and services in real time, enabling proactive and efficient responses to changing conditions [13–15].
- Real-time monitoring helps detect anomalies, identify potential issues, and optimize resource allocation. For instance, intelligent traffic management systems can reduce congestion by modifying signal timings based on current traffic circumstances.

9.2.3.3 Enhanced efficiency and resource optimization

- IoT technologies enable cities to optimize the use of resources, including energy, water, transportation, and waste management.
- By utilizing IoT analytics and data, cities can identify inefficiencies, implement intelligent resource allocation strategies, and reduce waste, leading to cost savings and improved environmental sustainability.

9.2.3.4 Improved citizen services and quality of life

- IoT-enabled services and applications improve the quality of life and citizen experience in smart cities.
- For instance, real-time information on parking availability is provided by smart parking systems, reducing the amount of time spent looking for space and reducing traffic congestion.
- Smart lighting systems adjust lighting levels based on occupancy, improving energy efficiency, and ensuring citizen safety.

9.2.3.5 Sustainable and environmentally friendly practices

- IoT facilitates using renewable energy sources in combination, optimizing energy generation, distribution, and consumption.
- IoT-based environmental monitoring helps track air quality, noise levels, and pollution levels, enabling cities to take appropriate measures to reduce environmental risks and improve the well-being of residents.

9.2.3.6 Citizen engagement and participation

- IoT technologies promote public participation in decision-making and citizen engagement.
- Through IoT-enabled platforms and software, citizens can provide feedback, access information, and actively contribute to the betterment of their city, fostering a sense of ownership and community involvement.

9.2.3.7 Safety and security

- IoT-based surveillance systems and emergency response mechanisms enhance safety and security in smart cities.
- Video analytics, sensors, and connectivity enable real-time monitoring, detection of anomalies, and timely response to security threats or emergencies, improving public safety and reducing crime rates.

9.2.4 Applications of IoT in smart city infrastructure

- **Smart traffic management systems:** IoT-enabled traffic management systems utilize data from different sensors, cameras, and interconnected vehicles to optimize traffic flow, reduce congestion, and enhance road safety.
- **Smart parking solutions:** IoT-based parking solutions save time spent looking for parking spaces and traffic congestion by providing real-time information on parking availability.
- **Waste management and environmental monitoring:** IoT sensors can be used to optimize waste collection routes and monitor waste levels in bins. To guarantee a better urban environment, environmental monitoring keeps an eye on factors like noise levels, air quality, and other variables (Figure 9.3).

Figure 9.3 Smart city applications.

Figure 9.4 Smart house.

- **Efficient street lighting:** Smart street lighting systems manage lighting levels based on current/real-time data, optimizing energy consumption, and reducing costs.
- **Smart surveillance and security systems:** IoT-based surveillance systems [16] enable remote monitoring of public areas, enhancing safety with security (Figure 9.4).

9.3 SMART HOMES

9.3.1 Introduction to IoT applications in the home environment

Smart homes are residential spaces equipped with Internet of Things (IoT) devices and systems that enable automation, control, and connectivity for various aspects of the home environment. These technologies enhance convenience, comfort, energy efficiency, and security, providing home-owners with greater control and flexibility in managing their residences.

9.3.2 Key aspects and features of smart homes [17]

- **Automation and control:** Smart homes incorporate automation and control systems that enable the remote management and scheduling of various home devices and appliances. Through IoT-enabled devices and platforms, homeowners can control lighting, HVAC systems, security systems, entertainment systems, and more, either through dedicated applications or voice assistants.
- **Connectivity and integration:** Smart homes utilize IoT connectivity to establish a network of interconnected devices and systems within the household. This connectivity allows devices to speak with one another and exchange knowledge, enabling seamless integration and coordination of different functionalities.
- **Energy efficiency:** Energy-saving techniques are used in smart homes to maximize energy efficiency and minimize waste. Smart meters and IoT-enabled sensors track energy use and give homeowners access to real-time data, allowing them to make educated decisions and alter their energy usage accordingly. Automated HVAC and lighting systems can be managed based on occupancy or predetermined schedules, which helps save energy.
- **Home security and surveillance:** Smart homes enhance security through IoT-enabled security systems. These systems include features such as smart locks, doorbell cameras, motion sensors, and surveillance cameras that can be remotely accessed and controlled. Real-time property monitoring, notifications, and even remote visitor access are all available to homeowners.
- **Convenience and personalization:** Smart homes offer convenience through voice-controlled assistants and smart speakers, allowing homeowners to control various aspects of their homes using voice commands. Additionally, they can create personalized settings and preferences for different devices and systems, such as customized lighting scenes, temperature presets, or personalized entertainment options.
- **Safety and environmental monitoring:** Smart houses can monitor environmental factors like air quality, temperature, and humidity as well as potentially dangerous substances like smoke or carbon monoxide. In the event of any anomalies or safety issues, home-owners are informed and given the opportunity to take urgent action.
- **Home entertainment and multimedia:** Smart homes provide enhanced multimedia experiences through integrated entertainment systems.

IoT-enabled devices can stream music, movies, and TV shows from online platforms, and users can control their entertainment systems through voice commands or smartphone applications.

- **Integration with wearable devices**: Smart homes can integrate with wearable devices such as fitness trackers or smartwatches. This integration enables the synchronization of data and the automation of certain home functions based on user activity or health metrics. For example, adjusting lighting and temperature settings when the user wakes up or arrives home.
- **Remote monitoring and management**: Smart home systems allow remote monitoring and management through smartphone applications or web interfaces. Homeowners can access and control various devices and systems even when they are far, providing calmness of mind and convenience.

9.3.3 Examples of smart home applications

- **Automated lighting and HVAC systems**: IoT-based solutions reduce energy waste by adjusting lighting and temperature based on user preferences and occupancy.
- **Assistants using voice commands and smart speakers**: IoT devices like voice assistants enable voice commands to control various smart home devices and provide information and entertainment.
- **Systems for access control and smart locks**: IoT-enabled door locks allow remote access control and monitoring, providing convenience and enhanced security (Figure 9.5).

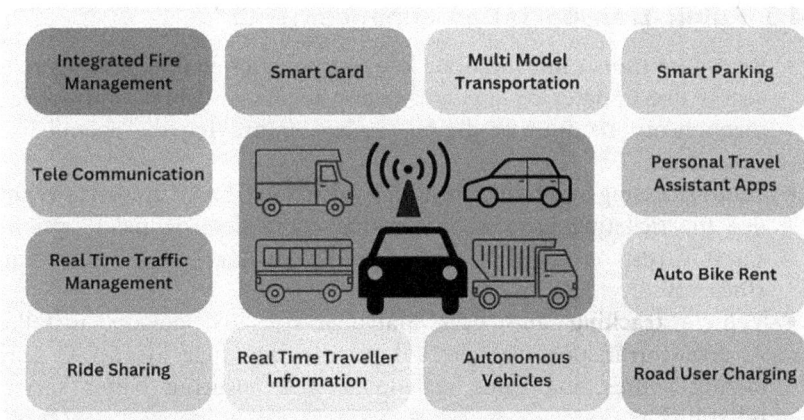

Figure 9.5 IoT applications in transportation and mobility.

9.4 SMART MOBILITY

9.4.1 Overview of IoT-enabled applications in transportation and mobility

The Internet of Things (IoT) has revolutionized the transportation and mobility sector by enabling many different applications that enhance efficiency, safety, and convenience. IoT-enabled technologies and devices are integrated into transportation systems, vehicles, and infrastructure to collect and analyze data, facilitate real-time communication, and better operations. Here are some important IoT-enabled applications in transportation and mobility.

9.4.2 Intelligent transportation systems (ITS)

- **Traffic flow optimization:** IoT sensors and cameras collect real-time data on traffic conditions, allowing transportation authorities to optimize traffic signal timings, reroute vehicles, and manage congestion more effectively.
- **Vehicles with connections:** IoT-enabled vehicles can communicate with each other and with transportation infrastructure, exchanging information about road conditions, hazards, and traffic patterns to improve safety and enable efficient traffic management.
- **Smart parking solutions:** IoT-based parking systems give drivers real-time information on parking availability, directing them to available spaces and easing traffic congestion brought on by cars driving about looking for parking.

9.4.3 Public transportation enhancements

- **Real-time transit information:** IoT technologies enable passengers to access live updates on bus and train schedules, delays, and arrivals through mobile apps or digital signage, improving the overall user experience.
- **Smart ticketing and fare collection systems:** IoT-based systems streamline the ticketing process by offering contactless payment options, smart cards, or mobile ticketing, reducing queues, and enhancing efficiency.
- **Vehicle tracking and fleet management:** IoT devices installed in public transport vehicles allow tracking their locations, optimizing routes, managing schedules, and ensuring better service reliability.

9.4.4 Smart supply chain management and logistics

- **Asset tracking**: IoT sensors and GPS technology enable real-time tracking of assets, such as vehicles, containers, or packages, ensuring efficient logistics operations, reducing theft, and improving supply chain visibility.
- **Inventory control**: IoT-based systems monitor inventory levels in warehouses and stores, providing real-time data on stock availability, optimizing supply chain operations, and reducing inventory holding costs.

9.4.5 Connected infrastructures

- **Smart traffic signals**: The total traffic control system is improved with IoT-enabled traffic signals that adjust in real time based on traffic flow.
- **Road condition monitoring**: IoT sensors embedded in the road surface provide data on temperature, moisture, and other parameters, allowing authorities to detect and respond quickly to adverse conditions such as icy roads or potholes.

9.4.6 Safety and security [10]

- **Vehicle and driver monitoring**: IoT devices, such as telematics systems, monitor vehicle performance, driver behavior, and safety metrics in real time, promoting safer driving practices and allowing preventive maintenance.
- **Emergency response systems**: IoT-based emergency response systems detect accidents or incidents and automatically trigger alerts or notifications to emergency services, improving response times and overall safety.

9.4.7 Ride-sharing and mobility services

IoT platforms and applications are utilized by ride-sharing companies and mobility service providers to match riders with drivers, manage vehicle fleets, track rides, and process payments seamlessly.

9.4.8 Unmanned vehicles

IoT plays a crucial role in autonomous vehicle technology, enabling real-time communication between vehicles, traffic infrastructure, and other elements of the transportation ecosystem. IoT sensors and connectivity facilitate the collection and analysis of data necessary for autonomous driving systems to operate safely and efficiently (Figure 9.6).

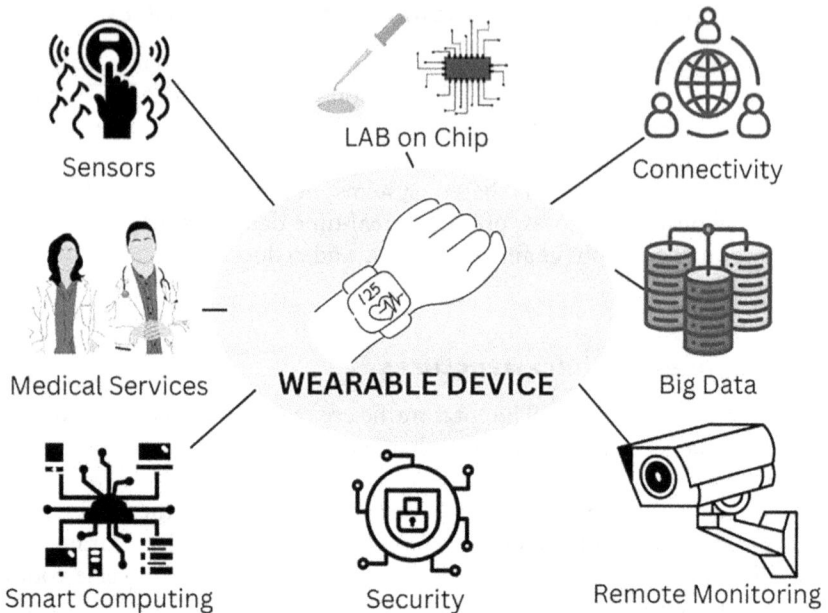

Figure 9.6 IoT applications in smart health.

9.5 SMART HEALTH

9.5.1 IoT applications in healthcare and wellness [18,19]

IoT applications in healthcare and wellness have revolutionized the way healthcare services are delivered, providing innovative solutions for remote patient monitoring, personalized care, and improved health outcomes. The integration of IoT devices, sensors, and platforms in healthcare systems has transformed the industry by enabling real-time data collection, analysis, and communication. Here are some key IoT applications in healthcare and wellness.

9.5.1.1 Online patient tracking

- **Wearable technology:** IoT-enabled wearables that track vital signs, activity levels, sleep patterns, and other health factors include smartwatches, fitness trackers, and biosensors. Real-time data transmission to healthcare practitioners enables remote monitoring of geriatric patients, people with chronic diseases, and people undergoing post-operative treatment.
- **Smart home monitoring:** IoT sensors integrated into homes can monitor the well-being and daily activities of individuals, providing support for independent living and enabling early detection of health issues or emergencies.

9.5.1.2 Telemedicine and virtual care

- **Virtual consultations:** IoT technologies enable remote video consultations between healthcare providers and patients, enhancing patient access to medical treatment, particularly in isolated or underdeveloped locations.
- **Remote diagnostics:** IoT devices, such as connected medical devices or home testing kits, allow patients to perform certain medical tests at home, with the results transmitted to healthcare professionals for remote diagnosis and treatment.

9.5.1.3 Medication management

- **Smart pill dispensers:** IoT-enabled pill dispensers check medication adherence, warn carers or healthcare providers when a dose is missed, and remind patients to take their meds at the appointed times.
- **Connected inhalers:** IoT inhalers monitor inhaler usage, provide usage reminders, and collect data on inhaler technique to help patients manage respiratory conditions effectively.

9.5.1.4 Personalized health and wellness

- **IoT fitness trackers:** Wearable fitness trackers gathering information on one's heart rate, activity level, sleep habits, and other health indicators, providing insights into personal wellness and enabling users to set and track health goals.
- **Smart scales and blood pressure monitors:** IoT devices can track weight, body composition, and blood pressure, providing personalized health data that individuals can monitor and share with healthcare professionals for proactive management of their health.

9.5.1.5 Ambient assisted living

- **IoT-enabled home monitoring systems:** Sensors installed in homes can detect falls, monitor daily activities, and give emergency notifications, safeguarding the security and welfare of the elderly or people with impairments.
- **Remote caregiver support:** IoT solutions enable caregivers to remotely monitor and manage the health and safety of their loved ones, providing peace of mind and improving care coordination.

9.5.1.6 Health data analytics

- **IoT data integration and analysis:** IoT-generated health data can be aggregated and analyzed to identify patterns, trends, and anomalies, enabling population health management, disease surveillance, and predictive analytics for early intervention and improved health outcomes.

- **Real-time alerts and notifications:** IoT systems can generate alerts or notifications to healthcare providers or caregivers based on predefined health thresholds or abnormal health indicators, ensuring timely interventions, and reducing hospital readmissions.

9.5.1.7 Hospital efficiency and asset management

- **IoT-enabled hospital systems:** IoT devices and connectivity in hospitals can optimize patient flow, track medical equipment, manage inventories, and automate workflows, improving operational efficiency, reducing costs, and enhancing patient care.

IoT applications in healthcare and wellness offer numerous benefits, including remote monitoring, personalized care, improved patient outcomes, and increased efficiency in healthcare delivery. These technologies have the potential to transform the healthcare landscape by empowering patients, enhancing preventive care, and enabling more effective management of chronic conditions.

9.5.2 Smart healthcare infrastructure

- **Asset and inventory management:** IoT-based systems streamline inventory management in healthcare facilities, ensuring the availability of necessary equipment and supplies.
- **Patient flow optimization in hospitals:** IoT technologies help manage patient flow, reducing waiting times and improving overall healthcare service delivery.

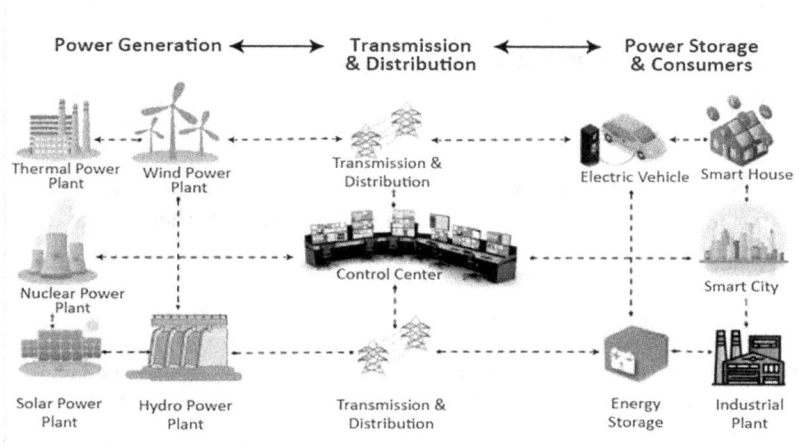

Figure 9.7 IoT in the energy sector.

- **Ambient assisted living for elderly care:** IoT devices and sensors in homes provide support for independent living, enabling monitoring of daily activities, fall detection, and emergency response (Figure 9.7).

9.6 SMART GRID

9.6.1 Introduction to IoT in the energy sector

The integration of Internet of Things (IoT) technology has significantly improved and streamlined the energy sector, which is critical to our daily life. The utilization of linked devices, sensors, and systems in the energy sector allows for the optimization of energy production, distribution, consumption, and management.

9.6.2 Smart grid infrastructure and IoT integration

The phrase "smart grid infrastructure" describes the upgraded electrical grid system that integrates cutting-edge technologies to enhance the effectiveness, dependability, and sustainability of the energy grid. These technologies include sensors, communication networks, Internet of Things (IoT) devices, and data analytics. The generation, transmission, distribution, and use of electricity are all being revolutionized by the smart grid architecture. Here are the key components and features of smart grid infrastructure:

9.6.2.1 Advanced metering infrastructure (AMI)

- **Smart meters:** Smart meters are IoT-enabled devices installed at consumers' premises that measure and record electricity usage in real time. These meters enable bidirectional communication between consumers and utility providers, providing detailed consumption data for accurate billing and demand-side management.

9.6.2.2 Communication networks [20]

- **Advanced communication systems:** Smart grid infrastructure utilizes robust communication networks, such as wired or wireless technologies, to enable seamless data exchange and real-time communication between different components of the grid, including smart meters, substations, and control centers.
- **Two-way communication:** The smart grid enables communication that is two-way between utility companies and customers, allowing for remote monitoring, control, and management of electricity supply and demand.

9.6.2.3 Grid monitoring and control

- **Sensor technology:** Smart grid infrastructure incorporates IoT sensors and devices throughout the grid to monitor and gather information on numerous characteristics, such as voltage, current, frequency, and power quality. These sensors provide real-time insights into grid conditions, enabling proactive maintenance, fault detection, and load balancing.
- **Distribution automation:** Smart grid technologies automate distribution processes, allowing remote monitoring and control of distribution equipment, fault detection and isolation, and quick restoration of power supply. This improves the reliability and resilience of the grid.

9.6.2.4 Grid optimization and energy management

- **Demand response:** Demand response programs, which reward consumers for adjusting their electricity use during periods of peak demand, are made possible by smart grid infrastructure. Consumers receive real-time pricing information and can optimize their consumption patterns accordingly, reducing strain on the grid.
- **Grid optimization:** IoT-enabled data analytics and predictive algorithms analyze real-time and historical data to optimize grid operations, load balancing, and power flow management. This improves the efficiency of energy distribution, reduces transmission losses, and ensures better utilization of grid assets.
- **Energy storage integration:** Smart grid infrastructure enables the integration of energy storage systems, such as batteries, into the grid. This allows for efficient storage and utilization of excess energy generated from renewable sources, providing grid stability, and reducing reliance on conventional power plants.

9.6.2.5 Renewable energy integration

- **Distributed energy resources (DERs):** The seamless integration of distributed energy resources, such as solar panels, wind turbines, and energy storage systems, is made possible by smart grid technology. IoT technologies enable real-time vigilance, regulation, and improvement of DERs, enabling better integration of renewable energy into the grid and promoting a cleaner and more sustainable energy ecosystem.

9.6.2.6 Data privacy and cybersecurity

To guard against online attacks and guarantee the privacy and integrity of data transmitted throughout the grid, smart grid infrastructure places a high priority on strong cybersecurity measures. Encryption, authentication, and secure communication protocols are employed to safeguard critical infrastructure and customer information.

Smart grid infrastructure brings numerous benefits [16,21], including enhanced grid reliability, improved energy efficiency, optimized energy management, integration of renewable energy sources, and increased engagement and empowerment of consumers. It lays the foundation for a more sustainable, resilient, and responsive energy ecosystem that supports the transition towards a greener future.

9.6.3 Energy monitoring and management

- Real-time monitoring of energy demand and consumption is made possible by IoT technology, both at the grid level and at the level of the individual user.
- Smart meters and sensors collect data on energy usage, peak demand periods, and power quality, providing valuable insights for energy management and optimization.

9.6.4 Demand response programs

- Demand response programs use communication networks and IoT-enabled smart meters to entice users to modify their energy use during spikes in demand.
- With access to real-time pricing information, consumers can limit their energy use during peak hours and contribute to a balance between supply and demand.

9.6.5 Renewable energy integration and optimization

- IoT plays a crucial contribution to integration and maximize the use of renewable energy, such as solar and wind power, into the existing energy infrastructure.
- IoT devices and systems monitor the performance of renewable energy installations, manage energy storage solutions, and enabling effective integration of renewable energy sources into the grid.

9.6.6 Energy efficiency and conservation

- By giving real-time data on energy consumption trends and enabling automatic control of energy-consuming equipment, IoT technology support energy saving projects.
- By intelligently controlling lighting, heating, ventilation, and air conditioning (HVAC) systems depending on occupancy and environmental

conditions, smart home and building automation systems use IoT to optimize energy usage.

9.6.7 Forecasting maintenance and asset management

- IoT devices and gadgets/sensors are deployed in energy infrastructure, such as power plants and distribution networks, to monitor equipment performance, detect anomalies, and enable predictive maintenance.
- Real-time monitoring and analysis of equipment data enable timely maintenance interventions, reduce downtime, and optimize asset management strategies.

9.6.8 Energy trading and grid optimization

- IoT technologies enable peer-to-peer energy trading platforms, where consumers with distributed energy resources (DERs) can directly buy and sell excess energy to other consumers.
- IoT-based energy management systems and analytics optimize grid operations by balancing energy supply and demand, improving grid stability, and minimizing transmission losses.

9.6.9 Data analytics and optimization

- Data generated by IoT paired with sophisticated analytics and machine learning techniques, provides insights for energy providers to optimize energy generation, distribution, and load management.
- Data analytics helps identify patterns, forecast demand, optimize energy resources, and enable proactive decision-making in the sector of energy.

IoT in the energy sector brings numerous benefits, including enhanced grid efficiency, improved reliability, greater using renewable energy sources in combination, optimized energy management, and increased consumer engagement. These technologies are essential in the shift to a more sustainable and effective energy ecosystem.

9.7 CASE STUDIES AND REAL-WORLD EXAMPLES

9.7.1 Smart cities – Barcelona, Spain

- Barcelona has implemented various IoT applications to transform itself into a smart city. For instance, the city optimizes trash collection routes and lowers expenses by using IoT sensors to track waste levels in underground containers.

• Additionally, in order to help drivers find available spaces and ease traffic congestion, Barcelona has installed smart parking systems that employ IoT sensors to identify parking space availability in real time.

9.7.2 Smart agriculture [18] – Chiba, Japan

• Chiba Prefecture in Japan has embraced IoT in agriculture to enhance crop production. Farmers employ IoT sensors to keep an eye on the temperature, humidity, and soil moisture levels, enabling precision irrigation and maximizing resource usage.

• IoT-enabled drones that have cameras and sensors are also utilized for crop monitoring, pest detection, and crop health assessment.

9.7.3 Connected healthcare [16] – M/s Philips Healthcare

M/s Philips Healthcare has developed IoT-based solutions to enhance patient care. For example, their IntelliVue Guardian Solution uses IoT sensors to continuously monitor patients' vital signs, alerting healthcare providers in real time about any critical changes, enabling early intervention and improving patient outcomes.

9.7.4 Smart transportation – M/s Uber and M/s Lyft

Ride-sharing companies like M/s Uber and M/s Lyft leverage IoT technologies to connect drivers and riders efficiently. Their mobile applications use GPS, real-time traffic data, and IoT-enabled navigation systems to match riders with nearby drivers, optimize routes, and provide estimated arrival times.

9.7.5 Industrial IoT – General Electric (GE)

M/s GE has implemented Industrial IoT solutions in manufacturing facilities to improve efficiency and reduce downtime. Their Predix platform collects and analyzes real-time data from machines and equipment, enabling predictive maintenance, optimizing production processes, and reducing energy consumption.

9.7.6 Smart retail – M/s Amazon Go

M/s Amazon Go is a cashier-less retail concept that utilizes IoT technologies. Customers use a smartphone app to enter the store, and IoT sensors, cameras, and machine learning algorithms track the items they pick up. The customers' virtual cart is automatically updated, and they are charged when they leave the store, eliminating the need for traditional checkout processes.

9.7.7 Smart energy management [10] – M/s Enel

M/s Enel, an energy company, implemented IoT solutions to optimize energy management. They utilize meters with smart facilities and IoT-enabled systems to monitor energy consumption, detect anomalies, and provide real-time energy usage information to customers, promoting energy efficiency and conservation.

These real-world examples showcase the diverse applications of IoT and IoT networks across different sectors, including smart cities, agriculture, healthcare, transportation, manufacturing, retail, and energy management. These applications demonstrate how IoT technologies can enhance efficiency, improve decision-making, and create innovative solutions to address various challenges in different industries.

9.8 CHALLENGES

Even while cutting-edge IoT and IoT network applications have many advantages, they must overcome a number of obstacles in order to be successfully implemented. Here are some key challenges in the innovative applications of IoT and IoT networks.

9.8.1 Privacy and security

- Security breaches and privacy concerns are major challenges in IoT deployments. With a vast number of interconnected devices and networks, ensuring the confidentiality, reliability, and accessibility of data becomes crucial.
- IoT devices are potential targets for hackers because they frequently collect and communicate sensitive personal or commercial information. To safeguard IoT devices and data, strong security measures are required, such as encryption, authentication, and access controls.

9.8.2 Interoperability and standardization

- Interoperability challenges arise due to the proliferation of IoT devices and platforms from different vendors. Lack of standardized protocols and data formats can hinder seamless communication and data exchange between devices and systems.
- Standardization efforts are necessary to ensure compatibility, interoperability, and scalability across IoT deployments, enabling devices and systems to work together seamlessly.

9.8.3 Scalability and complexity

- IoT applications often involve large-scale deployments with numerous devices, sensors, and data streams. Managing and scaling such complex IoT ecosystems can be challenging.
- Designing and maintaining robust IoT architectures, data management systems, and communication networks that can handle the increasing volume of data and devices is crucial.

9.8.4 Power consumption and energy efficiency

- IoT devices often operate on limited power sources, such as batteries. Balancing the functionality and power consumption of IoT devices becomes important for their long-term operation.
- To increase the battery life of IoT devices and decrease energy consumption, energy-efficient designs, low-power communication protocols, and optimized power management approaches are crucial.

9.8.5 Data analytics and management

- The enormous volume of data generated by IoT devices creates difficulties for data storage, processing, and analysis. Scalable and effective data management and analytics solutions are needed to handle and analyze enormous amounts of data in real time.
- To gain actionable intelligence from IoT data and relevant insights, it is essential to use advanced analytics techniques like machine learning and AI.

9.8.6 Regulatory and ethical considerations

- IoT implementations frequently entail the gathering and processing of sensitive and private data, which raises questions regarding data ownership, privacy, and legal compliance.
- Complying with privacy and data protection laws, ensuring transparency, and establishing ethical guidelines for data usage are essential for building trust among users and stakeholders.

9.8.7 Reliability and resilience

- IoT applications are expected to operate reliably in various environments and conditions. System failures, connectivity issues, or disruptions in communication networks can impact the reliability and performance of IoT deployments.
- Building resilient IoT networks, implementing redundancy measures, and ensuring fail-safe mechanisms are essential to maintain continuous operation and minimize downtime.

Addressing these challenges requires collaborative efforts from technology providers, policy-makers, and industry stakeholders. Overcoming these challenges will enable the successful implementation of innovative IoT applications, fostering the growth of connected ecosystems and unlocking the full potential of IoT technologies.

9.9 CONCLUSION

In the coming decades, the Internet of Things will make the "smart world," in which everything is interconnected, a reality. Although Internet of Things (IoT) is one of the main ways to communicate the idea of ubiquitous computing, it is still not as well known as cloud computing. The future of the Internet of Things depends on the integration of real or physical systems. The IoT has the most pervasive issue with security, making it a vital area where we must take action to secure data or information that is contained on a single network. Security is a persistent concern for every system. Protecting an IoT infrastructure includes a number of key elements, such as device identity and authentication methods. Consequently, to address a number of IoT device security vulnerabilities, cryptographic requirements serve as the foundation for authentication mechanisms. This chapter covered the security issues with each layer and the necessity for new security protocols to be created. The final section looked at some of the well-known IoT and IoT cloud paradigm applications, such as healthcare, smart cities, smart grids, smart transportation, etc. Based on the foregoing, it can be said that the IoT environment is a rich research area, especially when it comes to the topic of integration with cloud computing, which presents fresh ways to manage smart services and applications.

REFERENCES

1. Mohammad AbdurRazzaque, Marija Milojevic-Jevric, Andrei Palade, and Siobhán Clarke, "Middleware for Internet of Things: A Survey", IEEE Internet of Things Journal, Vol. 3, No. 1, February 2016.
2. Mayuri A. Bhabad, and Sudhir T. Bagade, "Internet of Things: Architecture, Security Issues and Counter measures", International Journal of Computer Applications, Vol. 17(4), pp. 2347–2376, 2015.
3. Cyril Cecchinel, Matthieu Jimenez, Sebastien Mosser, and Michel Riveill, "An Architecture to Support the Collection of Big Data in the Internet of Things", Services (SERVICES), 2014 IEEE World Congress on, pp. 442–449, 2014.
4. Rwan Mahmoud, Tasneem Yousuf, Fadi Aloul, and Imran Zualkernan, "Internet of Things (IoT) Security: Current Status, Challenges and Prospective Measures", in 10th International Conference for Internet Technology and Secured Transactions, 2015.

5. Ala Al-Fuqaha, Mohsen Guizani, and Mehdi Mohammadi, "Internet of Things: A Survey and Enabling Technologies, Protocols and Application", IEEE Communication Surveys & Tutorials, Vol. 17, No. 4, Fourth Quarter 2015.

6. Alessio Botta, Walter de Donato, Valerio Persico, and Antonio Pescape, "On the Integration of Cloud Computing and Internet of Things", 2014 International Conference on Future Internet of Things and Cloud (FiCloud), pp. 23–30, 2014.

7. Biplob R. Ray, Jemal Abawajy, and Morshed Chowdhury, "Scalable RFID Security Framework and Protocol Supporting Internet of Things", Computer Networks, Vol. 67, pp. 89–103, 2014.

8. H. Schaffers, N. Komninos, M. Pallot, B. Trousse, M. Nilsson, and A. Oliveira, "Smart Cities and the Future Internet: Towards Cooperation Frameworks for Open Innovation", The Future Internet, Lecture Notes in Computer Science, Vol. 6656, pp. 431–446, 2011.

9. Xu Xiaohui, "Study on Security Problems and Key Technologies of the Internet of Things", International Conference on Computational and Information Sciences, 2013.

10. Huansheng Ning, and Hong Liu, "Cyber-Physical-Social Based Security Architecture for Future Internet of Things", Advances in Internet of Things, Vol. 2(01), pp. 1–7, 2012.

11. Jayavardhana Gubbi, Rajkumar Buyya, Slaven Marusic, and Marimuthu Palaniswami, "Internet of Things (IoT): A Vision, Architectural Elements, and Future Directions", Future Generation Computer Systems, Vol. 29(7), pp. 1645–1660, 2013.

12. P. F. Harald Sundmaeker, P. Guillemin, and S. Woelfflé, "Vision and Challenges for Realising the Internet of Things", Pub. Office, EU, 2010.

13. Huansheng Ning, Unit and Ubiquitous Internet of Things. CRC Press, 2013.

14. Xiang Sheng, Jian Tang, Xuejie Xiao, and Guoliang Xue, "Sensing as a Service: Challenges, Solutions and Future Directions", IEEE Sensors Journal, Vol. 13(10), pp. 3733–3741, 2013.

15. Xue Yang, Zhihua Li, Zhenmin Geng, and Haitao Zhang, "Internet of Things", International Workshop, IOT 2012, Changsha, China, August 2012.

16. Rolf H. Weber, and Romana Weber, Internet of Things. Springer, pp. 41–68, 2010.

17. Ming Zhou, and Yan Ma, "QoS-Aware Computational Method for IoT Composite Service", The Journal of China Universities of Posts and Telecommunications, Vol. 20, pp. 35–39, 2013.

18. L. Atzori, A. Iera, and G. Morabito, "The Internet of Things: A Survey", Computer Networks, Vol. 54, No. 15, pp. 2787–2805, Oct. 2010.

19. Daoliang Li, and Yingyi Chen, Computer and Computing Technologies in Agriculture. Springer, pp. 24–31, Oct. 2010.

20. D. Giusto, A. Iera, G. Morabito, and L. Atzori (Eds.), Objects Communication Behavior on Multihomed Hybrid Ad Hoc Networks. Springer, pp. 3–11, 2010.

21. Wirelesssensornetwork, 2014. http://en.wikipedia.org/wiki/Wireless_sensor_network

Chapter 10

Taxonomy of botnet structure

Priyanka C Tikekar[1] and Swati S Sherekar[2]

[1]Bharatiya Mahavidyalaya, Amravati, Maharashtra, India

[2]P.G. Dept of Computer Science, SGBAU, Amravati, Maharashtra, India

10.1 INTRODUCTION

Botnet has developed into a pain point again for Internet and cybersecurity. Botnets are networks of zombies that a bad person controls for their evil purposes. These actions consist of distributed denial-of-service (DDoS) attacks, click theft, phishing, spamming, the spread of malware, traffic sniffing, etc. [1,2]. A botnet is a collection of infected computers, often known as zombies or bots, that is remotely controlled by a person known as the bot-master. The computer program known as a bot is what attacks or hacks the machines. A range of so-called distribution methods, including fraud and compromised websites that deliver software through drive-by mechanisms, are employed to spread the infection [3,4]. Once infected, the device will appear to the legitimate user to function normally while, in reality, it is capable of carrying out malicious tasks for the botnet, who will utilize a control and command (C&C) server to issue commands to and collect data from the zombie. Some of the main malicious activities on the internet, such as DDoS assaults, mail spamming, brute-force attacks port scans, and others, are carried out by botnets. It is dangerous because a coordinated group of infected hosts are concentrating on a single target [5,6]. The entire network can be taken down by botnets in a short amount of time. Numerous methods have been created to locate the botnet and destroy it, but attackers have demonstrated that they are constantly one step ahead of these methods (Figure 10.1).

10.2 STEPS IN A BOTNET ATTACK

The attack scenario is shown in Figure 10.2. First, the system will get infected if it is connected to the internet by a click in the email, downloading files, browsing web pages, etc. then these infected computers are finding a C&C server to get further instructions for consistent action, such as attack target after identifying a C&C server. In the first step, a bot

DOI: 10.1201/9781003461418-10

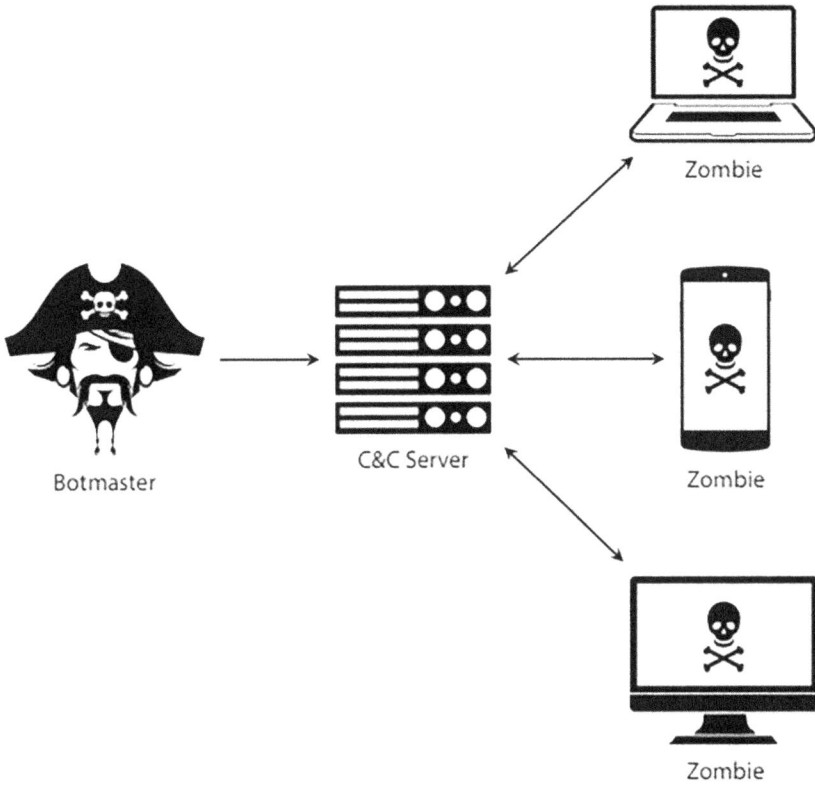

Figure 10.1 Botnet overview.

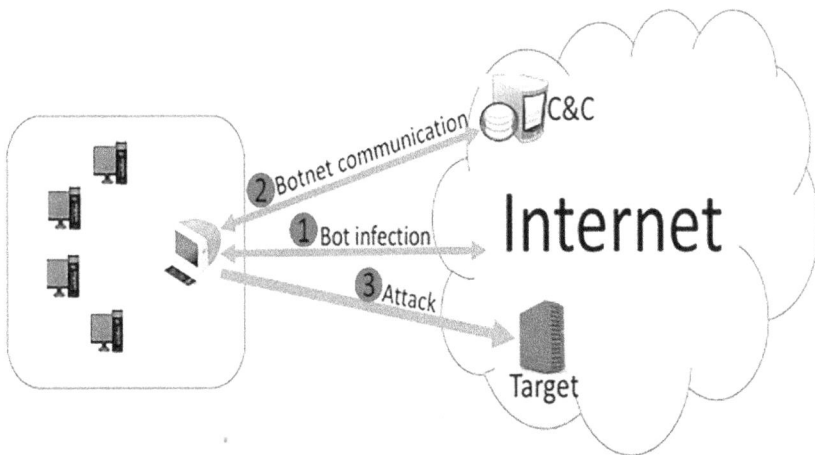

Figure 10.2 Typical steps in botnet attacks.

infection is spread in the system and then the botmaster sends commands to the various bots via a command-and-control channel and then at last the attack takes place, as shown in Figure 10.2 [7].

10.3 ARCHITECTURE OF BOTNET

Botnets are in general used for malicious purposes by the botmaster in order to disrupt the service operation. Botnet is a malware composed of different computers that are remotely controlled by a botmaster through a C&C server. A botmaster updates bots and controls the host action remotely. Botnet architecture is classified into three types: centralized botnet, distributed botnet, and hybrid botnet [8,9].

A. Centralized Botnet

In a centralized botnet, the botmaster sends commands to bots through a C&C channel. Bots are connected to a C&C server when they are active [10]. Communication is simplest in a centralized botnet; this topology uses a C&C server in which every bot connects directly to the server. The botmaster use a central server to issue commands to select bots. IRC and HTTP are the common protocols that are used in this centralized architecture. These two architectures are explained below (Figure 10.3) [11,12].

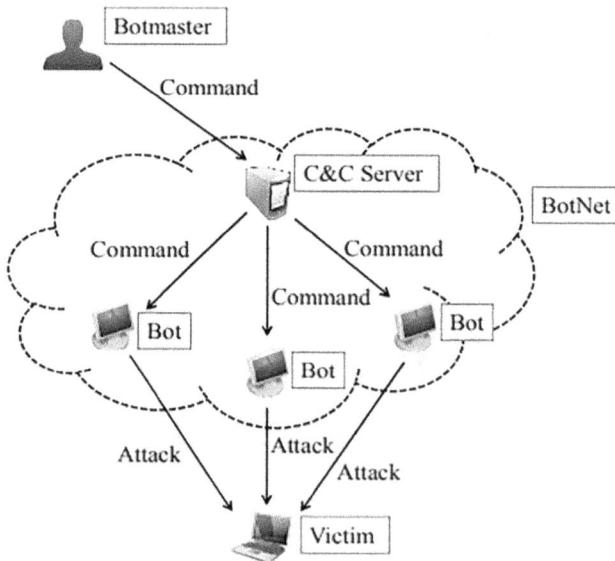

Figure 10.3 Centralized architecture.

a. IRC-Based Botnet

IRC stands for internet relay chat. It is a text-based instant messaging protocol. It is basically used for the communication between the client and server system. Lots of traffic is received by the IRC so it becomes easy to detect. A botmaster used an IRC channel for communication with the different bots. An IRC-based botnet works with the real-time online text messages and the botmaster uses this IRC due to its flexibility and simple structure. In this structure, the botmaster sends commands to different bots, also called zombies, that are connected in the network, which perform several malicious activities by establishing the IRC-based channel on the C&C server (Figure 10.4).

b. HTTP-Based Botnet

HTTP is the hypertext transfer protocol that is very popular while creating a botnet. With the help of the http botnet, botnet traffic can be easily hid by normal traffic. A botmaster sends the commands to the different bots, and it can easily bypass the firewall and hide malicious traffic with the normal

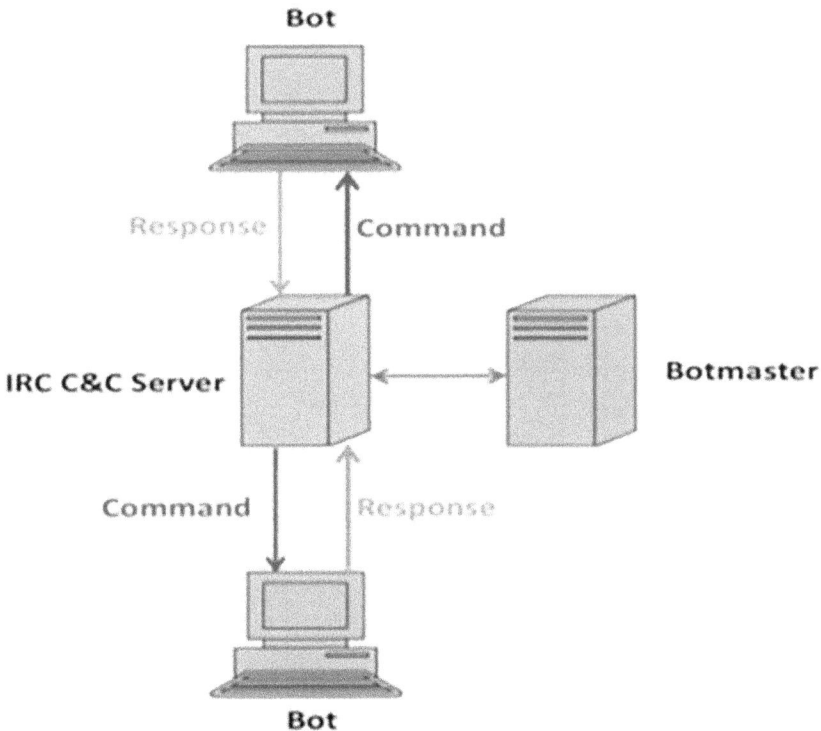

Figure 10.4 IRC-based botnet architecture.

http traffic. A HTTP-based botnet is more accurate than the IRC-based botnet. There is a problem of a central point of failure due to its centralized nature. When the botmaster uses legitimate websites, then in that case it will be difficult to detect the HTTP botnet (Figure 10.5) [13,14].

B. Distributed Architecture

Distributed architecture is also called decentralized botnets, sometimes referred to as P2P botnets. They don't have a centralized C&C channel. Bots can communicate via P2P protocol. In this topology, every bot is connected to at least one other bot. Only if every bot has the capacity to transmit commands to bots that are directly connected can orders be transmitted to the entire botnet [15,16]. It has no central server and each bot act as both client and server at the same time. This architecture is most commonly based on P2P protocols. One of the famous P2P botnets is the Strom botnet [8]. It uses an autonomously propagating virus to expand further in order to make use of networking services' vulnerability for remote code execution. In order to spread from one computer to another, if the attack is succeeding, its malware copies on its own and runs it on the victim's system (Figure 10.6) [17,18].

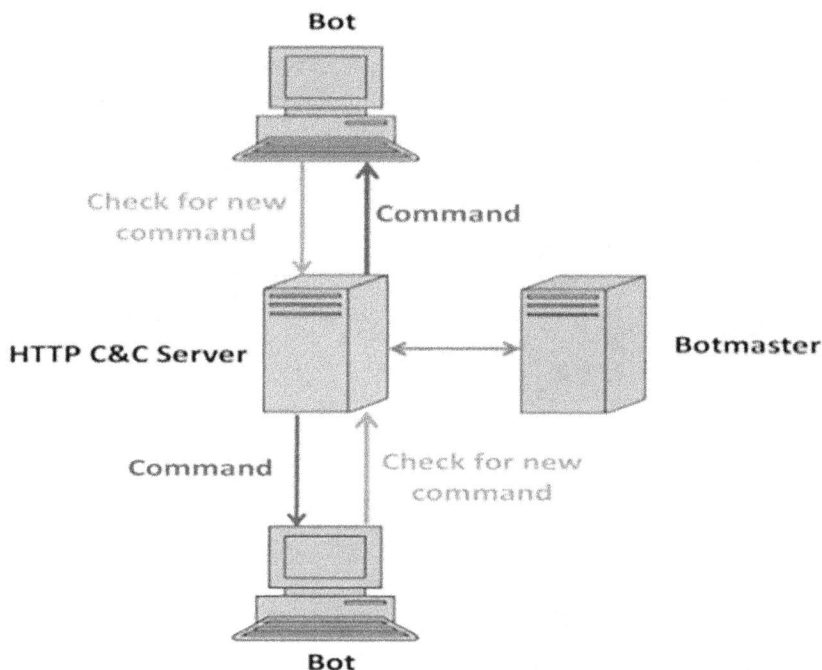

Figure 10.5 HTTP-based botnet.

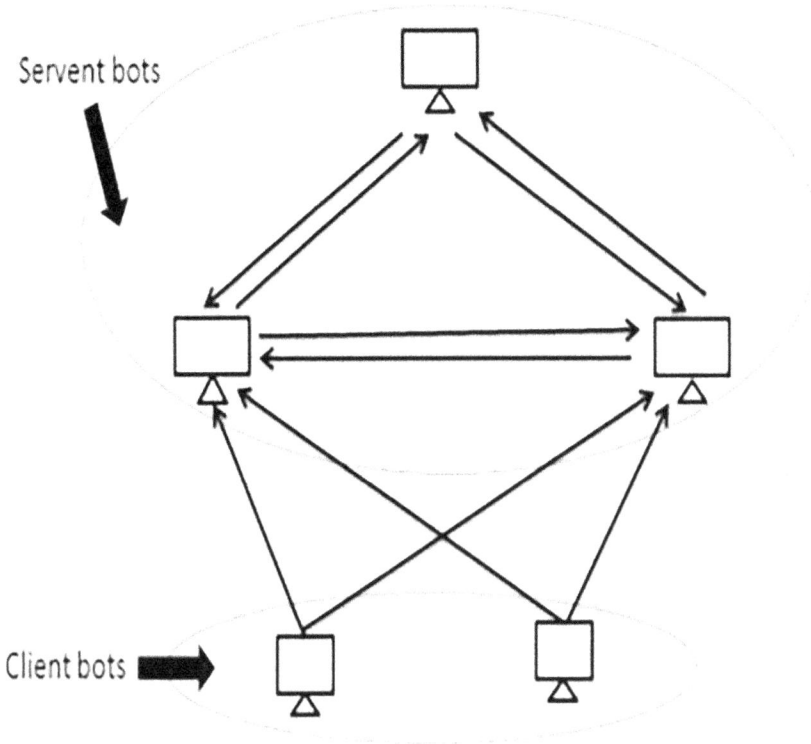

Servent bots

Client bots

Figure 10.6 Distributed architecture.

a. Peer-to-Peer Botnet

In the P2P-based botnet architecture, all the bots are connected with each other. The main objective of the P2P protocols is to hide C&C server. Various bots are used by the botmaster to issue the commands at every time. In this P2P structure, there is no central server, i.e., bots are dependent on the connected bots with-out having command and control server. Close to about 70% botnets are created using a peer-to-peer botnet. A botmaster uses any node to communicate with the other bots; it acts as a client server model in which each node can work as a client or server. The benefit of this P2P botnet is that a single bot can be detected but it does not mean that the whole botnet will be detected (Figure 10.7) [13,14].

C. Hybrid Architecture

Hybrid architecture is the combination of both centralized architecture and distributed architecture. The managing of P2P botnet topologies is complicated; hence, botnet fraudsters are going toward hybrid architecture, which

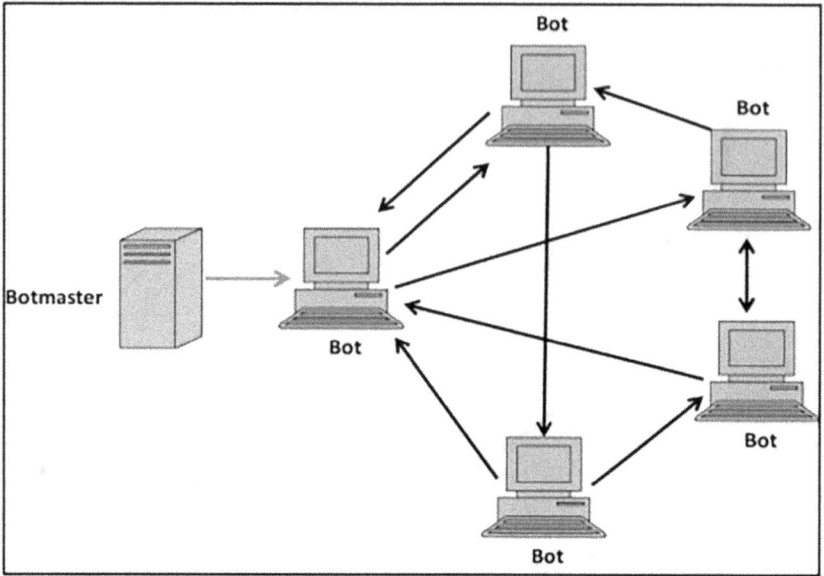

Figure 10.7 Peer-to-peer-based botnet.

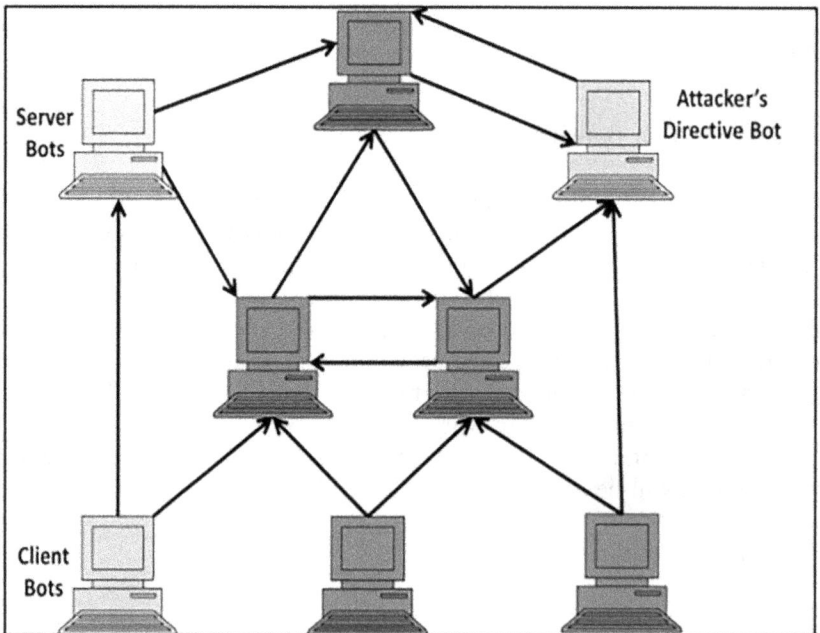

Figure 10.8 Hybrid architecture.

combines the advantages of both centralized and distributed architecture. One or more dispersed networks with one or more central servers each exist in a hybrid design. Spammers profit from this because, in the worst-case scenario, if one of these sites is disconnected, the other servers are not impacted, enabling the malware to remain operational as normal (Figure 10.8) [13,19].

10.4 BOTNET LIFE CYCLE

A botnet life cycle contains mainly five stages and starts with the initial infection. According to the life cycle, an attack is initiated by the botmaster, propagates through various stages, and then is finished. A susceptible device that is thought to be a potential bot is compromised during the life cycle's initial stage. The second stage involves downloading and installing the malware required to interact with a botmaster. To obtain instructions from the botmaster, the third stage involves connecting to the C&C server. The following level is the stage of harmful activity, which includes the infected host participating in malicious activities, as directed by the botmaster. The final phase entails maintenance and improvement (Figure 10.9) [20].

The botmaster needs to complete this stage in order to successfully monitor infected hosts for as long as feasible and alter their behavior by applying malware upgrades [21,22].

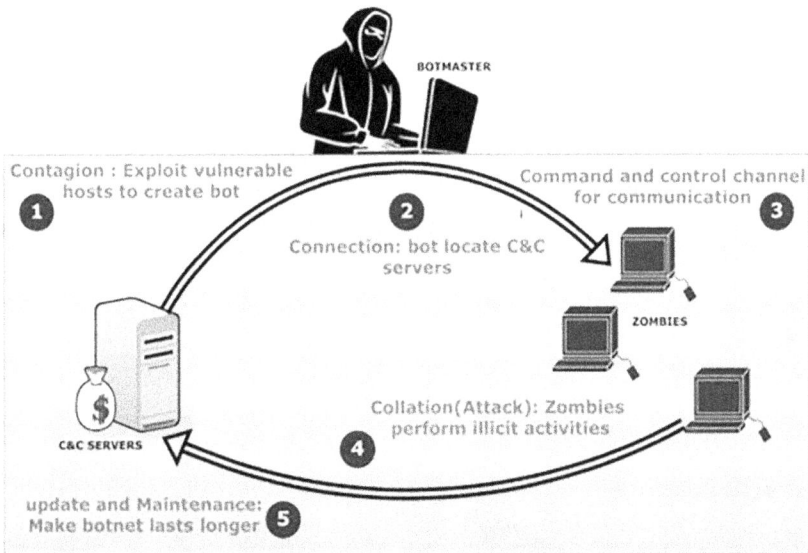

Figure 10.9 Life cycle of a botnet.

a. Initial Infection: In the initial infection phase, machines will get infected by downloading infected files or web downloads. The bot programs are forced to download once the susceptible computers have been located.

b. Secondary Injection: Whatever the program is installed in one first phase that will be executed in the second phase. When this code is run, it looks for and downloads bot binaries from the locations specified. The bot programmed is installed by the bot binary, which causes the weak machines to turn zombie. Every time a machine reboots, the bot begins running automatically.

c. Connection: In this phase, compromised machines tried to form a connection with the C&C channel. Once a C&C channel is established, machines will become an actual bot. It means that the connection is established in this phase.

d. Malicious Command & Control: As part of the C&C stage, the botmaster transmits instructions to start any illegal activity. The C&C server sends out the order.

e. Maintenance of Bots: The final stage, maintenance, will keep the bots active and current. The elimination of suspicious and dead bots from of the botnet is another issue it addresses [23].

10.5 TAXONOMY OF A BOTNET

The bot is the software application that will be executed through worms and other malicious code to perform certain cyber-criminal activities over the internet. A botnet is formed when the large number of bots are interconnected to each other (Figure 10.10).

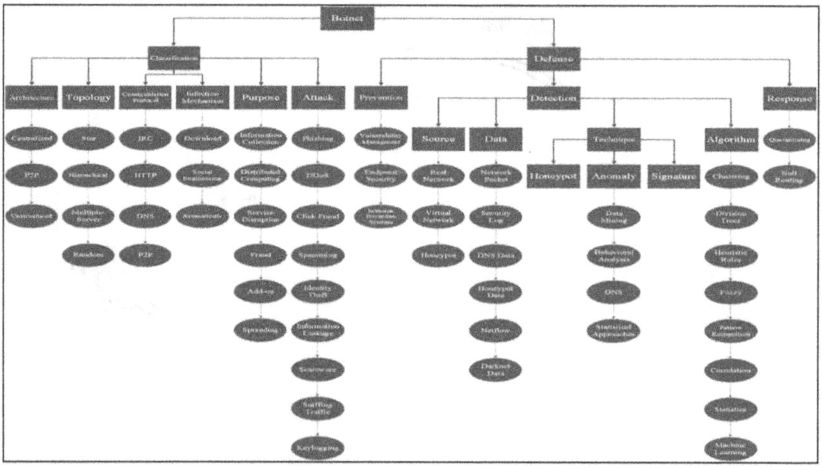

Figure 10.10 A new taxonomy of botnet.

There is one botmaster who will operate all the bots; it means that the botmaster is only responsible for sending the commands to different bots through which they can perform the malicious activities. In Figure 10.10, after covering all the previous or existing method, a new taxonomy on the botnet is discussed here. Here, classification is given and studies on the phenomenon of a botnet, its topology, architecture, protocols, purpose, and various attacks are given [24].

A. Classification

a. Architecture: A C&C architecture issues the commands to the botnet and receives back reports form the computer. There are three types of architecture: centralized, distributed, and hybrid architecture. Centralized architecture sends the commands to different bots through a C&C server. If the central server gets crashed, that means if it stops working in that case, distributed architecture is used. in the distributed architecture, any node can act as both a client and server. P2P, i.e., peer to peer, is the example of distributed architecture. A P2P structure is very difficult to detect due to its distributed nature.

b. Topology
Botnet topology consists of a network in which it is organized for the botnet communication. The topology of the botnet can be classified into various types, such as star, multiple server, hierarchical, and random. A star structure communicates with all the bots through a central server. When the botnet is used in the internet, then it forms the most popular structure, known as hierarchical. When a botmaster sends commands to a C&C server to perform attacks, these commands are broadcast to the different bots. Here multiple servers are used and it has the ability to enable and disable the server and client. In this topology, a botmaster sends commands to different bots through a C&C server but a random botnet does not have a C&C server.

c. Communication Protocol
Here, various communication protocols such as IRC, HTTP, P2P, and DNS are used while classifying the botnet. IRC is the internet relay chat protocol and HTTP is the hypertext transfer protocol. These two protocols are basically belonging to the centralized architecture. P2P is the peer-to-peer protocol that basically belongs to the distributed architecture. A botnet also uses the DNS, i.e., domain name system, protocol for the communication channel. To overcome the drawback of centralized architecture, the attacker uses a P2P protocol, which is used in distributed architecture.

d. Infection Mechanism

Infection can be spread by downloading the email attachment, web downloading, etc. Social engineering is used for spreading the malicious program. In the automatic bot, automatically scanning, exploitation, distributed computing, frauds, etc., a botnet attack occurs when the bots are infected and then they will perform various malicious activities such as click frauds, steal personal data of the user, identity theft, spamming, etc.

e. Attacks

There are various types of attacks, classified into phishing, DDoS, click fraud, identity theft, information leakage, scareware, sniffing, and keylogger. For performing a phishing attack, the attacker uses social engineering sites for exploring the vulnerabilities. There are the popular DDoS bots, such as Ago-Bot, SD-Bot, R-Bot, Spy-Bot, etc. For financial gain purposes, a click fraud attack is used. A Zeus bot is mainly focused on identity theft. Scareware is a fraud s/w that is usually covers the security of the software, such as anti-malware application.

B. Defense

a. Prevention: To provide protection against the botnet attack. There are three approaches, such as vulnerability management, endpoint security, and intrusion prevention system. Vulnerability.

b. Detection: In this detection phase, emphasis is given on the resources and mechanism of botnet detection. Based on the network traffic, a botnet detection mechanism is proposed that has different characteristics.

c. Response: Quarantining and null routing are the two basic reactions against the botnet. Quarantining is used to separate and limit the movement of a system or computer. In a null route, it is a path to a real nowhere in the network [24].

10.6 TAXONOMY OF BOTNET BEHAVIOR

In this taxonomy, botnet behavioral features are classified into propagation, rallying, command and control, purpose, and evasion. Other features are also classified further in these high-level botnet features. Propagation is the important step in a botnet in which its main objective is to steadily increase the number of bots and make a network. Most of the bot binary includes inbuilt tools that will help them to spread to a new host. These transmission methods can be divided into two categories: active and passive (Figure 10.11).

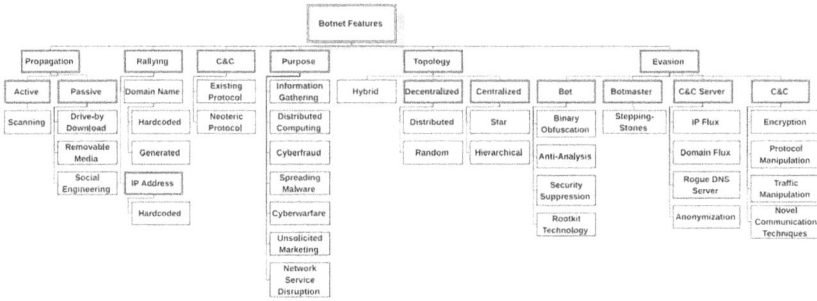

Figure 10.11 Taxonomy of botnet behavior.

Second is a rallying mechanism in the process used by bots in which a C&C server is discovered. There are some methods of a rallying mechanism that are commonly used, such as IP address, domain name, hard coded, generated, etc. A C&C plays a very important role in the botnet. Without a C&C server botnet, it is just a collection of infected machines. For carrying out command-and-control communication, there are several ways by which existing protocols are used. Existing protocols have been tried and tested and have less bugs compared with the custom protocols. Second, one is the neoteric protocol in C&C, a botnet uses of proprietary application-level protocol that is used for C&C communication. The main purpose of the botmaster is to manage hundreds and thousands of bots to combine to carry out the malicious activities. The main aim is the information gathering, distributed computing, cyber-fraud, spread malware, cyberwarfare, unsolicited marketing, network service disruption, etc. The next feature is the evasion, in that a botnet operates stealthily, enhance their probability, and duration of survival, etc. (Figure 10.12) [25].

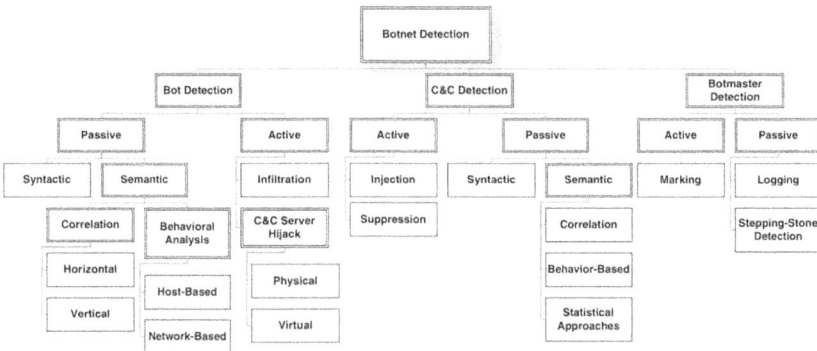

Figure 10.12 Taxonomy of botnet detection mechanisms.

10.7 TAXONOMY OF BOTNET DETECTION MECHANISM

Botnet detection is classified into three ways: bot detection, C&C detection, and botmaster detection, etc.

A. Bot Detection

Bot detection is done without concern to families of bots. in this detection, a bot can be detected by monitoring active and passive monitoring. Their primary aim is to prevent or protect the network from the infections without concerning bot families. Other than this, researchers are interested in identifying the bot families. In this bot detection, it indicates the vulnerability of a host or network to the botnet infection. In these two approaches, i.e., in active it involves the participation in the bot operations and in passive detection, this approach detects a botnet by observation and analysis of botnet activities. Passive techniques are syntactic and semantic. A syntactic approach is also known as a signature-based approach in which it identifies bots by comparing with predetermined patterns of the botnet infection, which are obtained from the observed samples. A semantic approach is used in the context of events and protocol for the detection of malicious behavior.

B. C&C Detection

In the botnet detection, C&C plays a very important role. Identification and analysis of C&C helps to understand botnet behavior. In the C&C detection, there are also two types of monitoring: active and passive. In the active detection, it takes part in the botnet operation. In the active monitoring, injection is used to inject the packets into suspicious network flows and suppression is used where incoming and outgoing traffic is suppressed in anomalous network flows; one can receive a measured response from each of the command-and-control communication's ends. The passive detection is used for C&C detection in addition to an active method. There are two passive approaches: syntactic and semantic. A signature-based model of C&C traffic is developed in syntactic detection. In the semantic approach, some heuristic approach is used for associating a certain behavior with C&C traffic. For the detection of C&C communication, some statistical approaches are also used for detection of a C&C botnet.

C. Botmaster Detection

A botmaster is the most important part of the botnet because a botmaster is only responsible for sending commands to different bots through a command-and-control channel. It is the most protected part of the botnet and that is why it will become difficult to detect. In this botmaster detection, there are again two approaches: active and passive. In the active botmaster detection, it involves manipulation of bot activities. In the passive detection of a botmaster, it involves the analysis of network traffic and other data, which will be without alteration or manipulating operations of a botnet. Routers log information about a packet that is passed through in the logging mechanism. Further, this information is used for the verification, whether suspected packets are forwarded to the specific routers or not [25].

10.8 CONCLUSION

Cybersecurity is the major issue in this era due to large multimedia data between the nodes in the network. There are lots of cybersecurity threats that are coming out lately so there is a need to overcome these threats. Among the various cybersecurity threats, a botnet is one of the most dangerous among them. It is the collection of a number of bots that are controlled by the botmaster and perform certain malicious activities like click frauds, identity theft, steal personal credentials, etc. Hence, the detection of bots and the botmaster in a cloud environment is important. Bot-affected nodes are identified using a trust-based evaluation approach. This approach consists of feature extraction and classification modules, as discussed here. In this chapter, the overview of a botnet, architecture of a botnet, its life cycle, and their taxonomy have been discussed.

REFERENCES

1. Ms. P. C. Tikekar, Dr. S. S. Sherekar, Dr. V. M. Thakre and Ms. A. S. Sherekar, "Comparative Analysis of Mobile Botnet Detection Techniques", *The National Conference on Emerging Trends in Science (NCETS)*, India, ISSN:2348-7143, 1–2 February 2019.
2. Ms. P. C. Tikekar, Dr. S. S. Sherekar and Dr. V. M. Thakre, "An Approach for P2P Based Botnet Detection Using Machine Learning", *Proceeding "Third International Conference on Intelligent Computing Instrumentation and Control Technologies (ICICICT)"*, IEEE ACCESS, pp. 1778–1791, 2022. doi: 197-1-6654-1004/22
3. Ms. P. C. Tikekar, Dr. S. S. Sherekar and Dr. V. M. Thakre, "A Study of Botnet Architecture & Its Defense Mechanism", *National Conference on Recents Advances in Science and Technology (AJANTA)*, India, ISSN: 2277-5730, 5–6 March 2019.
4. P. C. Tikekar, S. S. Sherekar and V. M. Thakre, "Features Representation of Botnet Detection Using Machine Learning Approaches", *International Conference on Computational Intelligence and Computing Applications (ICCICA)*, IEEE ACCESS, pp. 1–5, 2021. doi: 10.1109/ICCICA52458.2021.9697320

5. R. Abrantes, P. Mestre and A. Cunha, "Exploring Dataset Manipulation via Machine Learning for Botnet Traffic", *Procedia Computer Science*, vol. 196, pp. 133–141, 2022. ISSN 1877-0509. doi: 10.1016/j.procs.2021.11.082

6. Ms. P. C. Tikekar, Dr. S. S. Sherekar and Dr. V. M. Thakre, "Critical Analysis of Botnet Detection Techniques for Web Applications", *International Conference on Innovative Trends and Advances in Engineering and Technology (ICITAET)*, SHEGAON, IEEE ACCESS, pp. 89–93, 2019. doi: 10.1109/ICITAET47105.2019.9170246

7. Y. Jin, H. Ichise and K. Iida, "Design of Detecting Botnet Communication by Monitoring Direct Outbound DNS Queries", *IEEE 2nd International Conference on Cyber Security and Cloud Computing*, pp. 37–41, 2015. doi: 10.1109/CSCloud.2015.53

8. B. Choi, S.-K. Choi and K. Cho, "Detection of Mobile Botnet Using VPN", *Seventh International Conference on Innovative Mobile and Internet Services in Ubiquitous Computing*, pp. 142–148, 2013. doi: 10.1109/IMIS.2013.32

9. A. A. Santos, M. Nogueira and J. M. F. Moura, "A Stochastic Adaptive Model to Explore Mobile Botnet Dynamics", in *IEEE Communications Letters*, vol. 21, no. 4, pp. 753–756, 2017. doi: 10.1109/LCOMM.2016.263 7367

10. G. Vormayr, T. Zseby and J. Fabini, "Botnet Communication Patterns", in *IEEE Communications Surveys & Tutorials*, vol. 19, no. 4, pp. 2768–2796, 2017. Fourth quarter. doi: 10.1109/COMST.2017.2749442

11. A. Al-Ghushami, N. Karie and V. Kebande, "Detecting Centralized Architecture-Based Botnets Using Travelling Salesperson Non-Deterministic Polynomial-Hard Problem-TSP-NP Technique", *IEEE Conference on Application, Information and Network Security (AINS)*, pp. 77–81, 2019. doi: 10.1109/AINS47559.2019.8968710

12. H.-W. Hsiao, S.-H. Tung, M.-H. Shih and W.-P. Sung, "Using Botnet Structure to Construct the Communication System of a Real-Time Monitoring Platform: Botnet Structure for Real-Time Monitoring Platform", *13th International Conference on Natural Computation, Fuzzy Systems and Knowledge Discovery (ICNC-FSKD)*, pp. 2860–2865, 2017. doi: 10.1109/FSKD.2017.8393235

13. N. Kaur and M. Singh, "Botnet and Botnet Detection Techniques in Cyber Realm", *International Conference on Inventive Computation Technologies (ICICT)*, pp. 1–7, 2016. doi: 10.1109/INVENTIVE.2016.7830080

14. H. Xia, "Research on Bot-Net Prevention and Control Technology Based on P2P", *2nd IEEE Advanced Information Management, Communicates, Electronic and Automation Control Conference (IMCEC)*, pp. 1986–1990, 2018. doi: 10.1109/IMCEC.2018.8469569

15. M. Eslahi, R. Salleh and N. B. Anuar, "MoBots: A New Generation of Botnets on Mobile Devices and Networks", *International Symposium on Computer Applications and Industrial Electronics (ISCAIE)*, pp. 262–266, 2012. doi: 10.1109/ISCAIE.2012.6482109

16. H. R. Zeidanloo, M. J. Z. Shooshtari, P. V. Amoli, M. Safari and M. Zamani, "A Taxonomy of Botnet Detection Techniques", *3rd International Conference on Computer Science and Information Technology*, pp. 158–162, 2010. doi: 10.1109/ICCSIT.2010.5563555

17. W. Z. Khan, M. K. Khan, F. T. Bin Muhaya, M. Y. Aalsalem and H.-C. Chao, "A Comprehensive Study of Email Spam Botnet Detection", in *IEEE Communications Surveys & Tutorials*, vol. 17, no. 4, pp. 2271–2295, 2015. Fourth quarter. doi: 10.1109/COMST.2015.2459015

18. P. Wainerite and H. Kettani, "An Analysis of Botnet Models", in *the Proceeding of "3rd International Conference on Computer & Data Analysis"*, New York, NY, USA, pp. 121–166, 2016. doi: 10/1145/3314562

19. X. Dong, J. Hu and Y. Cui, "Overview of Botnet Detection Based on Machine Learning", *3rd International Conference on Mechanical, Control and Computer Engineering (ICMCCE)*, pp. 476–479, 2018. doi: 10.1109/ICMCCE.2018.00106

20. A. O. Prokofiev, Y. S. Smirnova and V. A. Surov, "A Method to Detect Internet of Things botnets", *IEEE Conference of Russian Young Researchers in Electrical and Electronic Engineering (EIConRus)*, pp. 105–108, 2018. doi: 10.1109/EIConRus.2018.8317041

21. S. Garg and R. M. Sharma, "Anatomy of Botnet on Application Layer: Mechanism and Mitigation", *2nd International Conference for Convergence in Technology (I2CT)*, pp. 1024–1029, 2017. doi: 10.1109/I2CT.2017.8226284

22. S. Bansal, M. Qaiser, S. Khatri and A. Bijalwan, "Botnet Forensics Framework: Is Your System a Bot", *Second International Conference on Advances in Computing and Communication Engineering*, pp. 535–540, 2015. doi: 10.1109/ICACCE.2015.124

23. H. Dhayal and J. Kumar, "Botnet and P2P Botnet Detection Strategies: A Review", *International Conference on Communication and Signal Processing (ICCSP)*, pp. 1077–1082, 2018. doi: 10.1109/ICCSP.2018.8524529

24. P. Amini, M. A. Araghizadeh and R. Azmi, "A Survey on Botnet: Classification, Detection and Defense", *International Electronics Symposium (IES)*, pp. 233–238, 2015. doi: 10.1109/ELECSYM.2015.7380847

25. S. Khattak, N. R. Ramay, K. R. Khan, A. A. Syed and S. A. Khayam, "A Taxonomy of Botnet Behavior, Detection, and Defense", in *IEEE Communications Surveys & Tutorials*, vol. 16, no. 2, pp. 898–924, 2014. Second Quarter. doi: 10.1109/SURV.2013.091213.00134

Chapter 11

A comprehensive study on a meta-heuristic optimization approach for maximum power point tracking in a solar power system

Zaiba Ishrat[1], Kunwar Babar Ali[2], and Taslima Ahmed[3]

[1]Department of Electronics & Communication Engineering, Meerut Institute of Technology, Meerut, Uttar Pradesh, India

[2]Department of Computer Science & Engineering (AI), Meerut Institute of Engineering & Technology, Meerut, Uttar Pradesh, India

[3]Department of Electronics & Communication Engineering, IIMT College of Engineering, Greater Noida, Uttar Pradesh, India

11.1 INTRODUCTION

Solar photovoltaic (PV) systems are regarded as more efficient than other renewable energy sources, due to their benefits to the environment and economy [1]. The cleanest, most abundant, least polluting, and least maintenance-required renewable energy source is solar electricity generation. The absence of moving parts and the absence of noise are the primary benefits of employing a solar PV system. Sun-derived solar irradiation is transformed to electrical power by means of a solar photovoltaic array. The sources of installed electrical energy in India as a percentage, by 30 June 2022, renewable energy sources would account for 28.25% of total energy capacity (MW). Share (%) reveals the following in which wind is 35.76%, solar is 50.59%, biomass is 8.95%, waste to energy is 0.42%, and small hydro is 4.29%. The photovoltaic cell operates on the electroluminescence effects, which convert light signals directly into electrical signals [2,3].

A solar panel system (SPS) consists of many photovoltaic panels that are linked in parallel and series to achieve a required rating. As a result, there is a strong likelihood that a partial shading condition (PSC) will develop. This problem arises when the SPS's surrounding environment and irradiance are not uniform. A universal peak and minor peaks are seen on the PV or PI plot of SPS under PSC. Numerous traditional MPPT algorithms have been developed [1] to improve SPS transferrable efficiency under uniform irradiance; however, these traditional procedures are unsuitable for tracking the maximum point under PSC. Therefore, a number of strategies for obtaining the highest power from SPS have been suggested in the literature [4–7].

DOI: 10.1201/9781003461418-11

In this review, the three most-used optimization algorithms are cuckoo search algorithm (CSOA), particle swarm optimization (PSOA), and artificial bee colony (ABCOA); they are brought together and their usefulness and tracking speed convergence to the MPP are evaluated. Section 11.2 discuss the algorithm overview, implementation in solar energy, and recent research in comparative analysis in terms of major findings and results of CSOA, PSOA and ABCOA, respectively. Section 11.3 discusses the comparative analysis of three swarm-based MPPT optimization techniques in terms of their merits and limitations and then finally a conclusion of the chapter, which provides a valuable resource for engineers, scientists, and professionals in the area of solar energy who are interested in the application of soft computing techniques.

11.2 OPTIMIZATION ALGORITHM FOR MPPT

11.2.1 Classification of optimization algorithms

All forms of PV systems are handled using meta-heuristic methods in dynamically changing weather circumstances. This study provides the method demonstration for various strategies that are practicable to execute, along with their benefits, drawbacks, and parameter-based characteristics. Additionally, the literature survey of these techniques contained will be a striking analysis for the specific application. Figure 11.1 categorizes all optimization-based MPPT strategies; the three techniques are explored and contrasted in this literature review.

11.2.1.1 Cuckoo search (CS) optimization algorithm CSOA

In 2009, Xin-She Yang and Suash Deb introduced the concept of CSOA [8,9]. The compel family sponging of some cuckoo species, which involves the cuckoos laying their eggs in the nests of host birds of other species, served as its model. There are typically three types of family sponging: intra-specific, obliging, and shell invasion. There are three idealized guidelines for CSOA, as per the sponging activities of cuckoos.

a. Every cuckoo deposits one egg at an instant in a shell that is picked at random. This random walk is modeled by levy's flight distribution function.
b. The best nest will produce the best eggs for the following generation. This is represented by a fitness function.
c. The number of nests that are open is set, and the hostess bird set a prospect that the quantity of progeny (set by a cuckoo) it finds is between 0 and 1. The host bird has the option of abandoning its nest or destroying cuckoo eggs if the cuckoo's progeny are found. In either case, given a fixed number of nests, a new nest will be produced with a probability of Px.

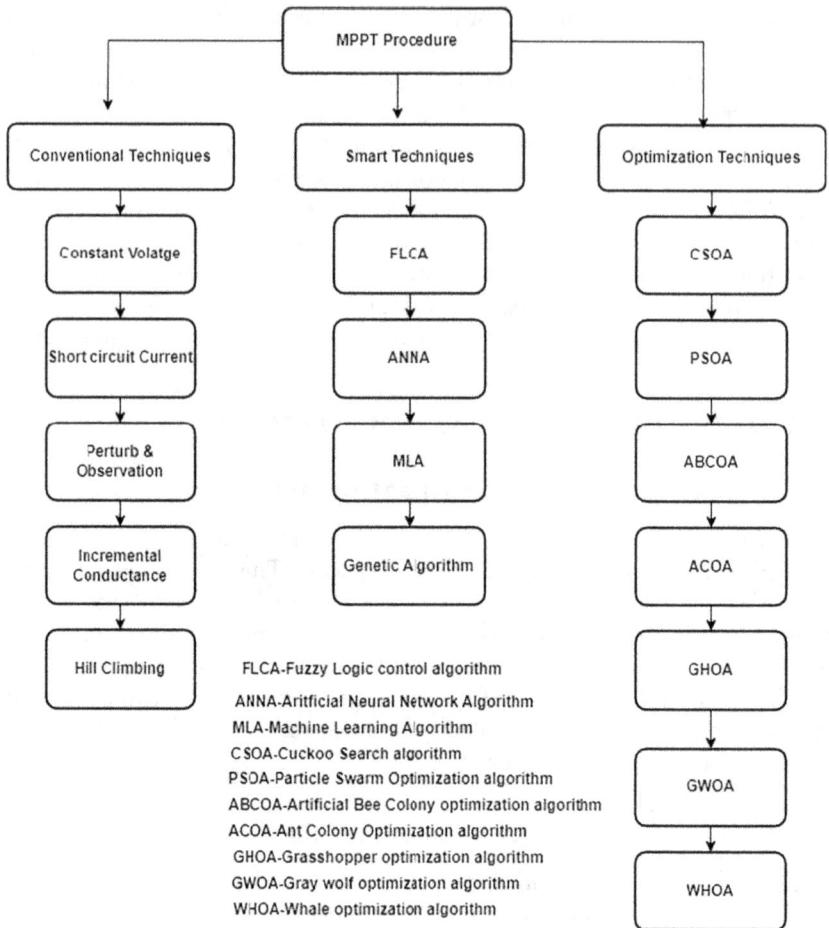

Figure 11.1 Classification of maximum point tracking algorithm.

The CSOA can be explained in a random code [8,9] by using these three rules. For the fitness function (FF), two variables, such as PV array operating voltage and step size Δ (for fine-tuning of GMPP), are used. It is given by Equations (11.1 and 11.2) [8,9]:

$$FF = f(V) \text{ and } Vt + 1 \tag{11.1}$$

$$i = Vi + \Delta \oplus \text{levy} \cdots (\lambda) \tag{11.2}$$

A flowchart of CSOA is shown in Figure 11.2 and Table 11.1 shows the comparison of CSOA MPPT approach span.

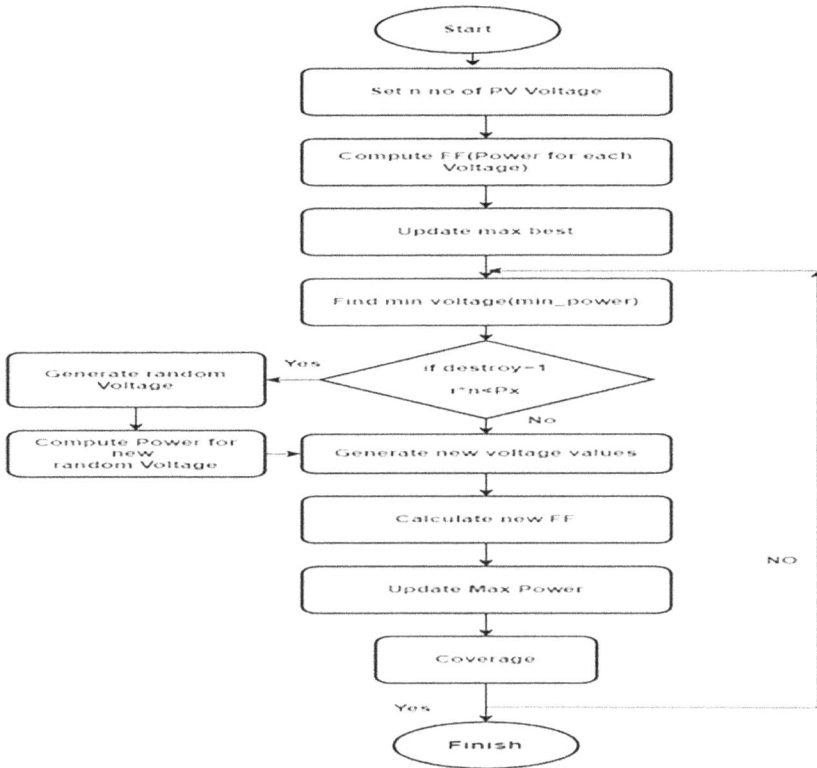

Figure 11.2 Flowchart of CSOA MPPT [9].

11.2.1.2 PSO particle swarm optimization (PSOA)

The fundamental tenet of PSOA is to utilize a population of particles, each of which represents a probable elucidation for the optimization issue. Every fleck's position is updated related to its current velocity, which is then restructured on the basis of an optimum situation of the other flecks in the population as well as the particle's own. A combination of the particle's current velocity, its ideal position, and the ideal positions of the other particles in the population is used to update each particle's velocity [17,18]. Consider a collection of Nm particles (Pi)2<i<Nm; a method is based on the following five steps [19]. Figure 11.3 displays the flowchart of PSOA and Table 11.2 shows the comparative analysis of previous work.

Step a: Haphazard situation of every fleck Pi by Equation (11.3):

$$Pi = \beta, \ 1 \leq i \leq Nm \tag{11.3}$$

where β is a haphazard number [Pinf ... Psup]

Step b: Every fleck discovers its neighborhood optimum place (Popti).

Table 11.1 Comparison of CSOA MPPT approaches from literature survey

Authors	Sense parameter	Hardware/ software	Convertor	Findings	Result
J. Ahmed et al. [10]	Power	Matlab/Simulink	Boost	Result compared by P&O method under PSC and shows zero oscillation around MPP under steady state.	Tracking Time (CS) 22 ms for STC and for P&O method 4, 5 ms.
J. Ahmed et al. [11]	Power	DSP- TMS320F2 8035)controller Matlab/Simulink	Buck-Boost	CSOA with GSS result compared with the PSO and CS method shows better tracking accuracy and less tracking time.	Tracking Accuracy = 99.77% ± 0. 28 Tracking time = 2.9.5 ± 0.44
D. A. Nugraha et al. [12]	Voltage and current	Matlab/Simulink	Buck-Boost	Comparing the outcome to the P&O and PSO tactics.	Tracking duration 100 to 250 ms and energy hammering in stable condition is 0.000008%
I. Mohamed et al. [13]	Voltage	Matlab/Simulink	Boost	Results are compared with incond and ANN method and shows better power at MPP under STC and varying weather conditions.	MPP Power = 60.4728 W
Basha et al. [14]	Voltage and current	Matlab/Simulink	SEPIC	The CSOA tracking technique uses the complete search space of IV plot to provide accurate global power point with rapid convergence.	MPP voltage 230 V for first irradiation pattern and for second pattern it is 250 V.
K. Farag et al. [15]	Voltage and current	Matlab/Simulink	Boost	To locate a more efficient global maximum power point, CSOA outperforms PSO.	CS efficiency 99.31% compared to 99.28% for PSO.

B. R. Peng et al. [15]	Voltage and current	Matlab/Simulink	Boost	The outcomes are contrasted with P&O tactic and variable P&O tactic and offers less number of iterations and optimal maximum point by easy computational calculation.	Comparing the traditional (P&O) and variable-step P&O algorithm, tracking duration was reduced by 46.42% and 11.76%, respectively.
Razak et al. [16]	Voltage and current	Matlab/Simulink	Boost converter	Compare the result of PSO and CS method with conventional technique and shows that CS performs better than other techniques under PSC.	Tracking time 0.24 s Tracking efficiency 99.87–99.94%

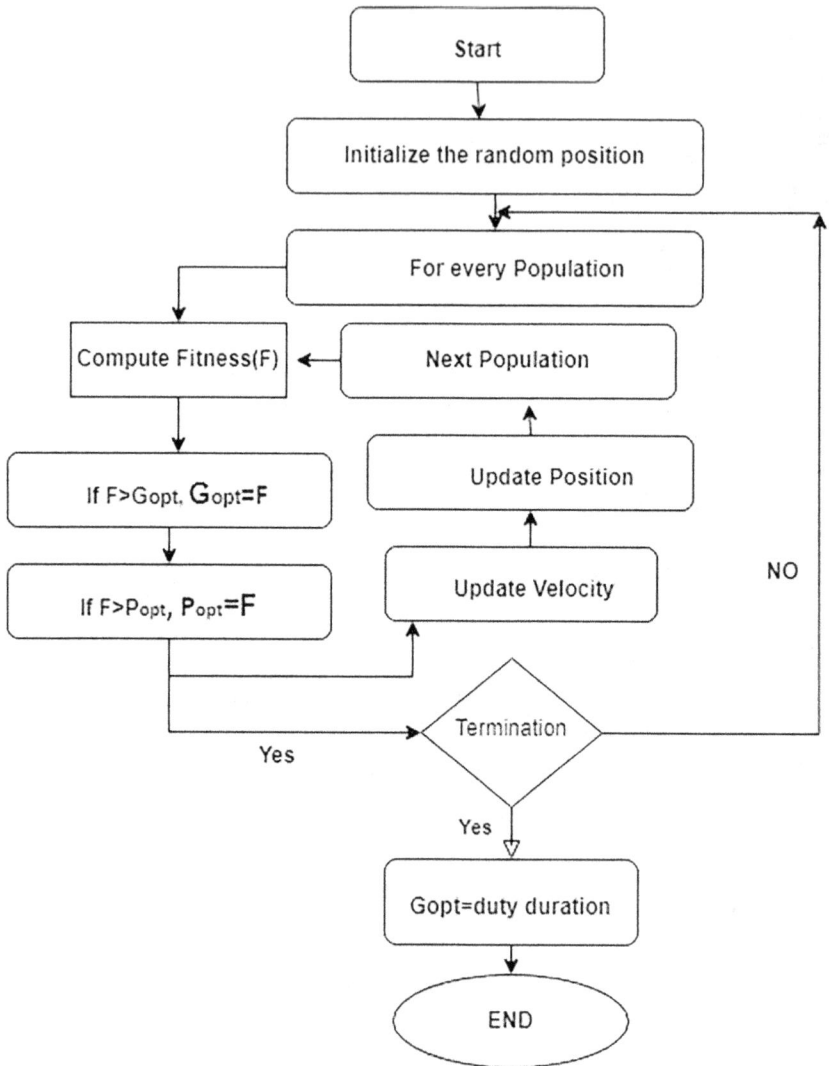

Figure 11.3 Flowchart of PSOA [19].

Step c: All flecks should track the global optimum place (PGopt).
Step d: Amendment of every fleck's place using Equations (11.4) and (11.5):

$$\Delta \text{Pi}^{(n+1)} = \Omega \times \Delta P_i^n + r1c1 (\text{Popti} - P^n) + r2c2 (\text{PGopt} - P^n) \quad (11.4)$$

$$P_i(n + 1) = P_i^n + \Delta P_i^{n+1} \quad (11.5)$$

Table 11.2 Comparison of PSOAMPPT approaches from previous works

MPPT technique	Sense parameter	Hardware/software	Convertor	Findings	Result
Hayder et al. [19]	Voltage and current	Matlab/Simulink	Boost	Result shows better performance compared to ANN-PSO-, and P-PSO method under varying solar Irradiance but temperature is constant.	Accuracy % in steady state Ass = 99.9980
O. Ben et al. [20]	Voltage and current	Matlab/Simulink	Buck	Result is compared with P&O and Fuzzy-TS method under different meteorological condition.	ηMPPT-99.30 ISE-10-6
Anoop et al. [21]	Current	Matlab/Simulink	Boost	PSO-OCC tactics is analyzed with classical PSOA.	The anticipated tactic can follow the global MPP precisely and rapidly
N. Kalaiarasi et al. [20]	Voltage and current	dspace 1104 controller Matlab/Simulink	Zsource Inverter	Result is compared with P&O method and shows less tracking time to MPP.	Settled time = 0.4 s
H. Li et al [22]	Voltage and current	Matlab/Simulink	Buck	Result is compared with firefly MPPT and PO-PSO method and shows better efficiency, less tracking time.	Pave - 134.17 W Efficiency-99.82% Tracking time- 0.210(s)
M. Merchaoui et al. [23]	Voltage and current	Matlab/Simulink	Boost	Nonlinear weight distribution with PSO result shows better tracking speed with comparison to linear weight distribution under PSC.	Avg. tracking speed < 0.75 s
Chang et al. [24]	Voltage and current	PIC18F8720 microcontroller Matlab/Simulink	Boost	A weight value is modified as per the gradient and modify in power of P-V plot.	Maximum average power is 35.32, 37.28, 45.55, and 64.73 and tracking speed 0.55, 0.98, 1.12, and 0.67 s.
Martinez et al. [25]	Voltage and current	Matlab/Simulink	Boost	Compared the simulation result of different PSO techniques with P&O method shows better performance.	Steady state average efficiency > 99.75%

where $P^{(n+1)}$ is the fresh fleck's place; P^n is the real place of fleck, $\Delta P^{(n+1)}$ is the perturb quantity to implement at the real place; ΔPi^n is the perturbation in the prior iteration; Ω is the inertia mass; r1 and r2 are haphazard variable in $[0,1]$; c1 is the cognitive constant; c2 is the public constant; PGopt is the global optimum place of the chief swarm flacks; Piopt is the neighborhood optimum place of the ith-flecks.

Step e: Reiterate b, c, and d until every fleck's place congregates to the PGopt.

For the SPS, the Pi is the duty duration that should be adjusted to attain the optimum output from SPS. For SPS, the metric Equation (11.6) should be

$$\Delta Pi^{(n+1)} = \Omega \times \Delta P_i^n + \mu(\text{Popti} - P_i^n) + \pounds(\text{PGopt} - P_i^n) \tag{11.6}$$

where $\pounds + \mu + \Omega = 1$. PGopt is the duty period that is related to the universal energy, while Popti is considered the duty value of i-th flecks, which is related to the minor optimum energy produced during the nth repetition.

11.2.1.3 Artificial bee colony (ABCOA)

This tactic discusses how bees search for nectar together. There are three categories of bees are employed bees, observer bees, and explorer bees [26]. Employed bees find foodstuff and then inform others about the location of food to nearby bees by doing a waggle dance. A foodstuff must be chosen by the spectator bee based on the information it has been given [27]. The ABC algorithm combines the confined finding method (by hired bees) and

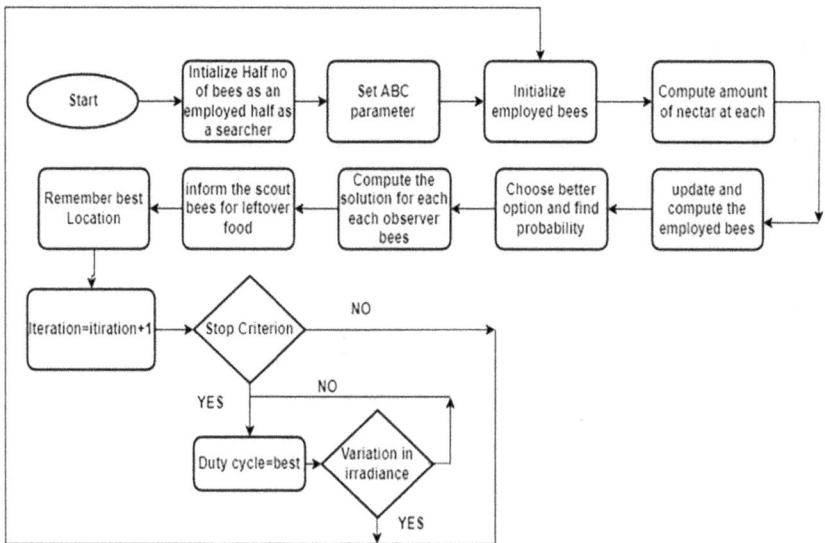

Figure 11.4 Flowchart of PSOA [19].

Table 11.3 Comparison of ABC optimization-based MPPT approaches from previous works

Authors	Sense parameter	Hardware/software	Convertor	Findings	Result
Catalina et al. [28]	Sun insolation and ambient temperature Vref is controlling parameter	PLECS RT Box 1 DSC (TI 28069 M)	Boost	The output voltage of SPS under variable weather conditions can be controlled by two sequential control loops; one is current (inner loop) controller and the other is voltage (outer loop) controller along with the ABCOA.	RE = 1.72 MAE = 0.4 RMSE = 0.42 Mean power = 39.92w Tracking factor = 98.92 Efficiency = 99.6%
Finani et al. [27]	Duty cycle Ppv	PSIM	Zeta	Human psychology optimization (HPO); procedure outcomes correlated with the ABCOA.	Tracking Accuracy = 99.95% Average Tracking Time = 0.0727 s
Sufiyan et al. [29]	Photon current, duty cycle	Matlab/Simulink and Cadence/PSpice	Boost	Comparison with PSO &PO MPPT	Accuracy = 98.76- 99.22% Convergence cycle = 10.4-10.1
M. Mao et al. [30]	Duty cycle	Matlab/Simulink	Buck Boost	Performance of P&O, PSOA, and ABCOA is compared with modified ABCOA.	Modified ABCOA shows better results in term of tracking efficiency.
Kanthala kshmi, et al. [31]	Voltage, current, duty cycle	Matlab/Simulink	Boost	ABC-PO algorithm with grid connected solar power system.	Efficiency = 99.93 Tracking speed = 0.08 s
Kanthala kshmi, et al. [32]	Voltage, current, duty cycle	Matlab/Simulink	Boost	ABC-PO algorithm for standalone system. Result is compared with ABC, PO, and INC method.	Efficiency > 99.5%

Table 11.4 Correlation table between CSOA, PSOA, and ABCOA

Parameter	CSOA	PSOA	ABCOA
Inspiration	Procreation behavior of cuckoo	Bird group trying to reach an unknown position	Foraging behavior of honey bee
Global optimum solution	Yes	Yes	No
Computational efficiency	Yes	Yes	More than CSOA and PSOA
Computational power need in RTOS	Less	More	Moderate
Parameter sensitivity	Yes	Yes	Moderate
Fitness function dependence	Yes	Yes	Yes
Local optimization stuck susceptibility	More	More	Less
Convergence speed	Moderate	Fast	Moderate
Gradient information need	No	Yes	Yes
Convergence dependency on initial condition	Yes	Yes	No [33,34]
Complexity	More	Simple	Simple
Control parameter	Less	More	Less

the comprehensive finding method to maintain a balance between exploration and exploitation (by onlooker and scouts). This is then applied to the PV system's MPPT by adjusting the duty cycle. Figure 11.4 displays the flowchart of ABCOA and Table 11.3 shows the comparative analysis of literature work (Table 11.4).

11.3 CONCLUSION

The three contemporary MPPT algorithms utilized in software and hardware platforms are briefly described in this review article. It deals with MPPT optimization methods that are mostly targeted in partial shading situations. In terms of SPS-MPPT tracking, each of the algorithms, CSOA, PSOA, and ABCOA, has benefits and shortcomings of its own. While the PSO method performs well globally, the CS approach is computationally economical and ideal for usage in embedded systems and real-time applications. The ABC algorithm is less prone to becoming stuck in local optima and is less sensitive to parameter selection. The MPPT tracking problem's specific requirements and the trade-offs between computing efficiency, parameter sensitivity, and optimization performance will determine which solution is best. The application, hardware accessibility, price, convergence speed, accuracy, and system dependability all affect the MPPT option that is selected. Given the significance of MPPT in partial shade conditions, it is clear that there is a large area of study to be done in order to identify an appropriate MPPT that can increase SPS output efficiency. This review is anticipated to be a very useful tool for all PV system researchers as well as for all sectors that excel at producing energy that is effective, clean, and sustainable for humankind.The authors declare that they have no conflict of interest related to this research work. Additionally, none of the researchers had any connections that could be seen as impacting the research's neutrality or the paper's content on a personal, professional, or financial level.

REFERENCES

1. Sahin, Z., Ismaila, K. G., Yilbas, B. S., and Al-Sharafi, A., "A review on the performance of photovoltaic/thermoelectric hybrid generators," *International Journal of Energy Research*, 44(5), pp. 3365–3394, 2020.
2. Central Electricity Authority (CEA), India. https://cea.nic. in/?lang=en (12 September 2022, date last accessed). Central Electricity Authority (CEA), India. Annual Report 2020–21 https://cea.nic.in/wpcontent/uploads/annual_reports/2021/CEAAnnualReport_fnal.pdf (15 August 2022 date last accessed).
3. Government of India, Ministry of Power. Power Sector at a Glance all India. https://powermin.gov.in/en/content/power-sector-glance-all-india (15 August 2022 date last accessed).
4. Farhat, M., Barambones, O., and Sbita, L., "Efficiency optimization of a DSP-based standalone PV system using a stable single input fuzzy logic controller," *Renewable and Sustainable Energy Reviews*, 49, pp. 907–920, 2015.
5. Sheik, S., Devaraj, D., and Imthias, T., "A novel hybrid Maximum Power Point Tracking Technique using Perturb & Observe algorithm and Learning Automata for solar PV system," *Energy*, 112, pp. 1096–1106, 2016.
6. Piegari, L., Rizzo, R., Spina, I., and Tricoli, P., "Optimized Adaptive Perturb and Observe Maximum Power Point Tracking Control for Photovoltaic Generation," *Energies*, 8, pp. 3418–3436, 2015.
7. Hassan, S. Z., Li, H., Kamal, T., Arifoglu, U., Mumtaz, S., and Khan, L., "Neuro-fuzzy wavelet based adaptive MT algorithm for photovoltaic systems," *Energies*, 10, p. 394, 2017.
8. Yang, X. S., and Deb, S., "Cuckoo search: State-of-the-art and opportunities," In: 2017 IEEE 4th International Conference on Soft Computing & Machine Intelligence, Port Louis, pp. 55–59, 2017.
9. Yang, X. S., and Deb, S., "Cuckoo Search via Levy flights," In: 2009 World Congress on Nature & Biologically Inspired Computing (NaBIC), Coimbatore, pp. 210–214, 2009.
10. Ahmed, J., and Salam, Z., "A soft computing MPPT for PV system based on cuckoo Search algorithm," In: 4th International Conference on Power Engineering, Energy and Electrical Drives, Istanbul, pp. 558–562, 2013.

11. Ahmed, J., and Salam, Z., "A Maximum Power Point Tracking (MPPT) for PV system using cuckoo search with partial shading capability," *Applied Energy*, 119, pp. 118–130, Apr. 2014.

12. Nugraha, D. A., Lian, K. L., and Suwarno, "A novel MPPT method based on cuckoo search algorithm and golden section search algorithm for partially shaded PV system," *Canadian Journal of Electrical and Computer Engineering*, 42(3), pp. 173–182, 2019.

13. Mosaad, M. I., abed el-Raouf, M. O., Al-Ahmar, M. A., and Banakher, F. A., Maximum power point tracking of PV system based cuckoo search algorithm; review and comparison, *Energy Procedia*, 162, 117–126.

14. Hussaian Basha, C. H., Bansal, V., and Rani, C. "Development of cuckoo Search MPPT algorithm for partially shaded solar PV SEPIC converter." https://link.springer.com/chapter/10.1007/978-981-15-0035-0_59

15. Abo-Elyousr, F. K., Abdelshafy, A. M., and Abdelaziz, A. Y., "MPPT-based particle swarm and cuckoo search algorithms for PV systems," *Modern Maximum Power Point Tracking Techniques for Photovoltaic Energy Systems, Green Energy and Technology*, doi: 10.1007/978-3-030-05578-3_14, Springer Nature Switzerland AG, 2020.

16. Rezk, H., Fathy, A., and Abdelaziz, A. Y., "A comparison of different global MPPT techniques based on meta-heuristic algorithms for photovoltaic system subjected to partial shading conditions," *Renewable and Sustainable Energy Reviews*, 74, pp. 377–386, July 2017, doi: 10.1016/j.rser.2017.02.051.

17. Díaz Martínez, D., Trujillo Codorniu, R., Giral, R., and Vázquez Seisdedos, L., "Evaluation of particle swarm optimization techniques applied to maximum power point tracking in photovoltaic systems," *Internation Journal of Circuit Theory Applications*, pp. 1–19, 2021.

18. Pilakkat, D., Kanthalakshmi, S., and Navaneethan, S., "A comprehensive review of swarm optimization algorithms for MPPT controller of PV systems under partially shaded conditions," *Electronics*, 24, pp. 3–14, 2020.

19. Hayder, W., Ogliari, E., Dolara, A., Abid, A., Hamed, M. B., and Sbita, L., "Improved PSO: A comparative study in MPPT algorithm for PV system control under partial shading conditions," *Energies*, 13, p. 2035, 2020, doi: 10.3390/en13082035.

20. Ben Belghith, O., Sbita, L., and Bettaher, F., "MPPT design using PSO technique for photovoltaic system control comparing to fuzzy logic and P&O controllers," *Energy and Power Engineering*, 8, pp. 349–366, 2016, ISSN Online: 1947-3818 ISSN Print: 1949-243X.

21. Anoop, K., and Nandakumar, M., "A novel maximum power point tracking method based on particle swarm optimization combined with one cycle control," In: Proceedings of International Conference on Power, Instrumentation, Control and Computing (PICC), Thrissur, India, pp. 1–6, Jan. 2018.

22. Li, H., and Yang, D., "An overall distribution particle swarm optimization MPPT algorithm for photovoltaic system under partial shading," *IEEE Transactions on Industrial Electronics*, 66(1), pp. 265–275, Apr. 2018.

23. Merchaoui, M., Saklyl, A., and Mimouni, M. F., "Improved fast particle swarm optimization based PV MPPT," In: Proceedings of the 9th International Renewable Energy Congress (IREC 2018), Hammamet, Tunisia, pp. 1–6, Apr. 2018.

24. Chang, L.-Y., Chung, Y.-N., Chao, K.-H. et al., "Smart global maximum power point tracking controller of photovoltaic module arrays," *Energies*, 11(3), pp. 567–582, Mar. 2018.
25. Kalaiarasi, N., Dash, S. S., Padmanaban, S. et al., "Maximum power point tracking implementation by dspace controller integrated through z-source inverter using particle swarm optimization technique for photovoltaic applications," *Applied Science*, 8(1), pp. 145–162, Jan. 2018.
26. Bilal, B., "Implementation of artificial bee colony algorithm on maximum power point tracking for PV modules," In: 8th International Symposium on Advanced Topics in Electrical Engineering ATEE 2013, pp. 1–4, 2013.
27. Fanani, M. R., Sudiharto, I., and Ferdiansyah, I., "Implementation of maximum power point tracking on PV system using artificial bee colony algorithm," In: 3rd International Seminar on Res. Inf. Technol. Intell. Syst. International seminar on research of information technology and intelligent systems (ISRITI) 2020, pp. 117–122, 2020.
28. Gonzalez-Castano, C., Restrepo, C., Kouro, S., and Rodriguez, J., "MPPT algorithm based on artificial bee colony for PV system," *IEEE Access*, 9, pp. 43121–43133, 2021.
29. Soufyane Benyoucef, A., Chouder, A., Kara, K., Silvestre, S., and Sahed, O. A., "Artificial bee colony based algorithm for maximum power point tracking (MPPT) for PV systems operating under partial shaded conditions," *Applied Soft Computing*, 32, pp. 38–48, 2015.
30. Nie, L., Mao, M., Wan, Y., Cui, L., Zhou, L., and Zhang, Q., "Maximum power point tracking control based on modified abc algorithm for shaded PV system," In: 2019 AEIT International Conference on Electrical and Electronic Technologies for Automotive, AEIT Automot 2019, 2019.
31. Pilakkat, D., and Kanthalakshmi, S., "Single phase PV system operating under partially shaded conditions with ABC-PO as MPPT algorithm for grid connected applications," *Energy Reports*, 6, pp. 1910–1921, 2020.
32. Pilakkat, D., and Kanthalakshmi, S., "An improved P&O algorithm integrated with artificial bee colony for photovoltaic systems under partial shading conditions," *Solar Energy*, 178 (November 2018), pp. 37–47, 2019.
33. Karaboga, D., "An idea based on honey bee swarm for numerical optimization," Technical Report TR06, Erciyes University, no. TR06, p. 10, 2005, doi: citeulike-article- id:6592152.
34. Sundareswaran, K., Sankar, P., Nayak, P. S. R., Simon, S. P., and Palani, S., "Enhanced energy output from a PV system under partial shaded conditions through artificial bee colony," *IEEE Transactions on Sustainable Energy*, 6(1), pp. 198–209, 2015, doi: 10.1109/TSTE.2014.2363521.

Chapter 12

Blockchain in ubiquitous computing

Ramander Singh[1], Ashish Pandey[2], and Lav Kumar Dixit[1]

[1]Department Computer Science & Engineering, RD Engineering College, Ghaziabad, Uttar Pradesh, India

[2]Department Computer Science & Engineering, IMS Engineering College, Ghaziabad, Uttar Pradesh, India

12.1 INTRODUCTION

In today's interconnected world, where technology is seamlessly integrated into our surroundings, the marriage of blockchain and ubiquitous computing offers a transformative paradigm. Blockchain, known for its secure and decentralized data management, finds a natural partner in ubiquitous computing, which envisions an environment rich with intelligent devices and services.

Blockchain's unalterable and transparent ledger, combined with its decentralized consensus mechanism, aligns well with the integrity and trust demands of ubiquitous systems. Imagine a future where smart cities seamlessly manage resources, IoT devices securely communicate, and wearable technology ensures personal data privacy – all powered by blockchain's immutable records.

12.1.1 Ubiquitous computing

A concept in computer science and engineering called ubiquitous computing, sometimes known as pervasive computing or ambient intelligence, imagines a day when computing capabilities are smoothly incorporated into regular items, settings, and activities. The objective of ubiquitous computing is to provide a setting in which technology is undetectable, unobtrusive, and context-aware, becoming a seamless component of people's lives without needing conscious awareness or direct user engagement.

In the late 1980s, Mark Weiser, a computer scientist at Xerox PARC (Palo Alto Research Center Incorporated), initially proposed the idea of ubiquitous computing [1]. In his worldview, computer devices would be integrated into daily life activities, facilitating organic interactions and improving user experiences. A foundation for ubiquitous computing is the growth of computing from massive mainframe computers to personal computers, mobile devices, and finally the seamless incorporation of computing into the real world.

DOI: 10.1201/9781003461418-12

The following are some essential attributes and concepts of ubiquitous computing:

1. **Invisibility:** The goal of ubiquitous computing is to make technology invisible so that people do not need to actively engage with it or be aware of its existence. While supporting consumers' requirements, the technology should function in the background and not call attention to itself.
2. **Context awareness:** Ubiquitous computing devices are built to comprehend the user's surroundings, movements, preferences, and actions in relation to the context in which they are used. Devices may modify their behavior and services based on this contextual knowledge.
3. **Interconnected device:** In an environment where computers are pervasive, gadgets are linked through a variety of communication protocols, such as the Internet of Things (IoT). This makes it possible for devices to communicate and share data without any issues, resulting in a more smooth and integrated user experience.
4. **Ubiquitous connectivity:** A key component of ubiquitous computing is connectivity. The internet or other networks should be accessible to devices, allowing for constant data interchange and real-time updates. Ubiquitous computing strives to build smart environments by integrating computer resources into physical locations. These surroundings include, for example, smart workplaces, smart residences, and smart cities.
5. **User-centric design:** In ubiquitous computing, user-centric design is crucial. Designing technologies and apps that meet user demands and easily integrate into everyday activities is the main goal.
6. **Scalability and flexibility:** Ubiquitous computing systems must be both scalable to support a large number of connected devices and flexible enough to change to meet the demands of users and keep up with technological improvements.

Security and privacy are crucial factors in ubiquitous computing because of how much computers are becoming integrated into everyday life. Systems and devices must safeguard user data and provide a secure connection.

Numerous industries, including smart homes, healthcare, transportation, retail, manufacturing, and entertainment, use ubiquitous computing. It makes it possible to create cutting-edge software and services that improve user experiences, increase productivity, and provide insightful data.

As technology advances, ubiquitous computing will become a more important factor in determining how we interact with the digital world, transforming it into a seamless and pervasive aspect of our everyday lives.

12.1.1.1 Ubiquitous architecture

Data from the physical world is gathered via the ubiquitous sensors layer. This information might include everything from location and movement to temperature and humidity. Data is transported from the network layer to the context-aware computing layer. By analyzing the data, the context-aware computing layer can identify the user's context. The location, activities, and preferences of the user may be included in this context. The user interface layer tailors the information's presentation to the user's context. The user may interact with the environment via the apps layer (Figure 12.1).

The many levels of a ubiquitous architecture are described in further detail below:

Pervasive sensors: Pervasive sensors are low-cost, tiny sensors that are integrated into their surroundings. They are capable of gathering information on the environment's physical characteristics, including temperature, humidity, position, and movement.

Network: Data from ubiquitous sensors is transported through the network layer to the context-aware computing layer. Wired or wireless networks are also possible.

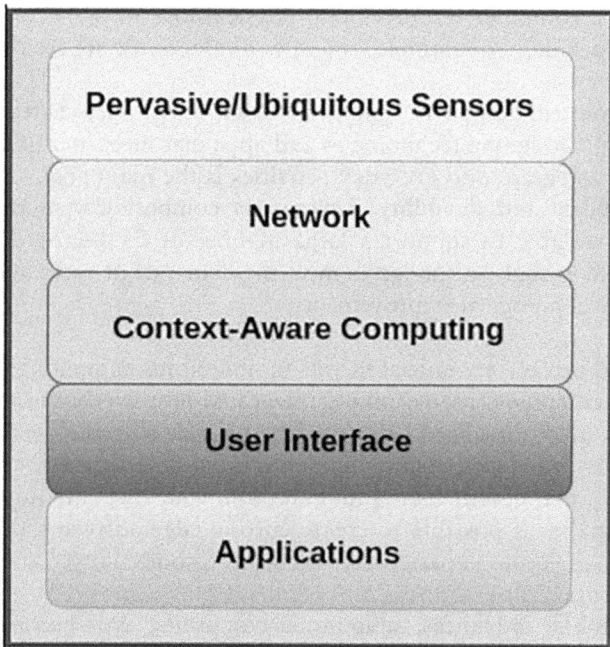

Figure 12.1 A typical ubiquitous architecture.

Computing that is aware of context: The context-aware computing layer analyzes data from omnipresent sensors to ascertain the user's context. The user's location, activities, and preferences might be included in this context.

User interface: The user interface layer gives the user information in a manner that is appropriate for their situation. Any number of gadgets, including a smartphone, tablet, or wearable device, may serve as the user interface.

Applications: The applications layer enables user interaction with the surrounding environment. Applications may be used in a number of contexts, including education, healthcare, entertainment, and productivity.

Despite being a complicated area, ubiquitous architecture has the power to completely alter how we interact with our surroundings. We may design a world that is more sensitive to our wants and preferences by integrating sensors into the environment and examining the data they gather.

12.1.2 Blockchain technology

The groundbreaking idea of blockchain technology has been quite well known since it was first proposed. Although it was first developed to serve as the foundation for cryptocurrencies like Bitcoin, its potential uses now go well beyond virtual money. A blockchain is fundamentally a distributed, decentralized ledger that securely logs transactions across a number of computers in a manner that assures its immutability, transparency, and security.

Here is a brief explanation of the main ideas behind blockchain technology:

1. **Decentralization:** To verify and log transactions, traditional systems often depend on a central authority. A blockchain, on the other hand, runs on a decentralized network of computers called nodes, each of which has a copy of the complete ledger. By doing so, the network's resiliency is improved and the necessity for a single central organization is removed.

2. **Distributed ledger:** A blockchain, which holds records of transactions that are arranged into blocks, is basically a digital ledger. A chronological chain of blocks is formed by each block, which includes a list of transactions, a timestamp, and a reference to the block before it.

3. **Consensus methods:** Consensus methods are used to preserve the integrity of the blockchain. These are the guidelines that govern the process of verifying and adding new transactions to the blockchain. Bitcoin uses the proof of work (PoW) consensus method, which forces nodes to solve challenging mathematical puzzles in order to verify transactions. Other approaches include delegated proof of stake (DPoS), practical Byzantine fault tolerance (PBFT), and proof of stake (PoS), each having its own benefits and drawbacks [2,3].

4. **Immutability:** It is very difficult to change or remove data from the blockchain after a block of transactions has been created. This is accomplished through cryptographic hashing, where each block has a distinct code (hash) based on its contents and the preceding block's hash. Due to the computational effort required, tampering is exceedingly rare since altering the contents in one block would necessitate changing the hash of all succeeding blocks.

5. **Protection:** A blockchain's decentralized structure, coupled with cryptographic methods, offers a high degree of protection against fraud and assaults. Because the network is decentralized, it is more difficult for malevolent parties to undermine the system because transactions are cryptographically signed to guarantee their validity.

6. **Smart contracts:** Blockchains may carry out self-executing contracts known as smart contracts in addition to transactions. These are programmed scripts that, when certain requirements are satisfied, instantly carry out an agreement's provisions. They may be used to simplify and automate many commercial operations.

7. **Public vs. private blockchains:** Public and private blockchains are the two primary varieties. Public blockchains are completely decentralized and accessible to everyone. Contrarily, private blockchains are only accessible to a small number of users and are often utilized in business situations where privacy and control are crucial.

8. **Use cases:** Although cryptocurrencies continue to be a prominent use of blockchain technology, its potential is broad. These include a variety of industries, such as real estate, voting systems, banking, healthcare, and supply chain management. Due of the transparency, traceability, and security that blockchain technology offers, other areas have begun exploring and experimenting with it.

Finally, the decentralized open and secure way of recording and authenticating transactions pioneered by blockchain technology. Its impact has extended well beyond its initial association with cryptocurrencies, opening doors for innovation in several other sectors.

12.1.3 Benefits for integrating blockchain and ubiquitous computing

There are a number of compelling reasons and advantages for integrating blockchain technology with ubiquitous computing, sometimes referred to as the Internet of Things (IoT), that may improve the efficiency, security, and reliability of both systems. The following are the main justifications for fusing blockchain with ubiquitous computing:

1. **Data integrity and immutability:** The volume of data generated and exchanged by IoT devices is enormous. The data produced by IoT

devices may be safely stored on the blockchain by using the immutability and cryptographic hashing of the blockchain. This guarantees the data's integrity and reliability, which is essential for key applications, including supply chain management, healthcare, and industrial operations.

2. **Decentralization and trust:** The dispersed nature of IoT networks and the decentralized design of blockchain are complementary. IoT devices may employ blockchain's consensus methods to build a decentralized trust model instead of depending on a central authority for data verification and authentication. This lowers the possibility of single points of failure and raises the system's overall dependability.

3. **Secure transactions and micropayments:** IoT devices often communicate with one another to carry out different functions. Smart contracts on the blockchain may help with safe, automated device-to-device transfers. This is especially helpful for allowing micro-transactions, in which even little sums of currency may be traded directly between devices.

4. **Identity and access management:** Blockchain can help IoT ecosystems with identity and access management. On the blockchain, identities of devices may be securely kept, and smart contracts can control access rights. This minimizes the danger of unauthorized access by ensuring that only authorized devices and users may interact with the network.

5. **Supply chain transparency:** By combining blockchain and IoT, supply chain management may be more transparent. Every stage of the supply chain journey may be recorded on the blockchain using sensors and RFID tags on items. This lowers the danger of product counterfeiting and enhances traceability by allowing stakeholders to confirm the authenticity and origin of items.

6. **Data monetization and ownership:** IoT device owners will have greater control over their data thanks to blockchain. While maintaining ownership, they might decide to share certain data with other parties. Smart contracts may enable new business models where users can monetize their data by facilitating safe data exchange and enforcing rules for data use.

7. **Standardization and interoperability:** By incorporating blockchain into the IoT ecosystem, interoperability issues may be resolved. Blockchain protocols may operate as a uniform layer to provide fluid communication and interaction amongst heterogeneous devices made by various manufacturers.

8. **Distributed analytics:** Analyzing the massive volumes of data produced by IoT devices may be resource-intensive. Blockchain may make it possible for data to be immediately analyzed on edge devices or across the network, enabling distributed analytics. This

might decrease the need for big data transfers to centralized servers, increasing efficiency.

9. **Auditing and compliance:** Blockchain's transparent and auditable nature may assist in guaranteeing that data is correctly recorded and readily verified, lowering the risk of fraud and mistakes in sectors that need tight auditing and compliance, such as healthcare and finance.

10. **Energy efficiency:** Compared to conventional proof of work (PoW) procedures, certain consensus techniques employed in blockchains, such as proof of stake (PoS), utilize less energy. This fits in nicely with the limited resources of IoT devices and makes the system as a whole more energy efficient.

Enabling new use cases and business models and combining blockchain with ubiquitous computing (IoT) may improve data integrity, security, trust, and transparency. The convergence of these technologies has the potential to transform a number of sectors and build stronger, decentralized, and more effective ecosystems.

12.2 CHALLENGES AND OPPORTUNITIES IN UBIQUITOUS COMPUTING

The Internet of Things (IoT), also known as ubiquitous computing, provides a variety of benefits as well as difficulties as it becomes more and more interwoven into our everyday lives and different businesses. Here are some main obstacles and chances related to ubiquitous computing.

12.2.1 Key challenges in ubiquitous computing

1. **Privacy and security:** IoT devices often connect to the internet and gather private information. This data's security and privacy must be guaranteed, which is a difficult task. Data leaks, breaches, and unauthorized access may all be caused by inadequate security measures.

2. **Data management and analytics:** Storing, processing, and analyzing the enormous amount of data produced by IoT devices is difficult. Scalable data collection, storage, and analysis techniques must be effective.

3. **Interoperability and standards:** The Internet of Things (IoT) landscape is made up of many devices from various manufacturers, each with their own set of communication standards and protocols [4]. For smooth communication between devices and systems, interoperability and standardization must be achieved.

4. **Scalability:** The networks and systems that enable IoT devices must be able to manage the scale as the number of these devices grows. System

performance might suffer due to congestion, latency, and scalability problems.

5. **Energy efficiency:** Many Internet of Things (IoT) devices rely on finite power sources, such as batteries. To increase device life spans and lessen the need for frequent battery changes, energy consumption must be optimized.

6. **IoT devices often function in real-time or mission-critical settings, which brings us to the next point:** reliability and quality of service. It may be difficult to maintain a high level of dependability and quality of service, particularly when dealing with connection problems or network congestion.

7. **Ethical and social implications:** The widespread use of computers creates moral concerns about who owns the data, who has the right to use it, and its possible social repercussions. It's crucial to strike a balance between technical development and moral issues.

12.2.2 Addressing security and privacy concerns

Considerations for security and privacy are essential when combining blockchain technology with ubiquitous computing. While both technologies have a lot to offer, protecting data's security, integrity, and availability comes with special difficulties. Considerations for security and privacy in the context of fusing blockchain with ubiquitous computing look like this:

12.2.2.1 Security challenges

1. **Device vulnerabilities:** IoT devices often have limited resources and may lack reliable security features. Due to this, they are vulnerable to several threats, including the introduction of malware, data eavesdropping, and unauthorized access.

2. **Blockchain vulnerabilities:** Despite the fact that blockchain is renowned for its security characteristics, there may still be flaws at different tiers, such as defects in smart contracts, weaknesses in the consensus process, and intrusions on the underlying infrastructure.

3. **Data tampering:** IoT devices provide data that is essential for corporate operations and decision-making. Maintaining the integrity of this data requires making sure it is not altered before it is transmitted to the blockchain.

4. **Physical attacks:** IoT devices may be physically accessed, which makes them vulnerable to theft or tampering. As a result, hostile actors could be able to access private data without authorization.

12.2.2.2 Privacy challenges

1. **Data leakage:** IoT devices may produce sensitive and private data. To preserve user privacy, it is essential to make sure that this data is not compromised during transmission, storage, or processing.
2. **Identifiable information:** IoT devices have the ability to gather information that might be used to identify specific people. It's difficult to strike a balance between the need of data collecting and the safeguarding of user identities.
3. **Data aggregation:** There is a chance that enough data may be gathered from various IoT devices to re-identify people or divulge private information.
4. **Public vs. private blockchains:** While public blockchains provide transparency, their openness may not be compatible with the privacy needs of certain use cases. Private blockchains provide users greater control over data visibility, but they also call for confidence in the parties involved.

12.2.3 Opportunities for enhancing interoperability and trust

Enhancing interoperability and trust among intricate and varied networks of linked devices is possible by incorporating blockchain technology into the field of ubiquitous computing (IoT). Here are several crucial chances to accomplish these objectives.

1. **Interoperability through standardization:**
   - For IoT devices to interact and conduct transactions, blockchain may function as a standardized layer. It may provide devices from many manufacturers a standard protocol to utilize, enabling smooth communication.
   - Smart contracts allow automatic interactions between machines that follow a preset set of rules, assuring uniform behavior across different kinds of machines.
2. **Decentralized identity and access management:**
   - Devices and users may have distinct identities that are confirmed by the blockchain thanks to blockchain-based decentralized identification solutions. As a result, centralized identification suppliers are no longer required.
   - Smart contracts built on the blockchain may be used to handle access control, ensuring that only approved devices and people have access to certain resources or data.

3. **Secure data sharing and monetization:**
   - Blockchain can make it possible for IoT devices to share data in a safe and auditable manner. Smart contracts allow data owners to determine the conditions of sharing [5].
   - With the use of this data sharing system, users may be able to monetize their data while still maintaining control over who has access to it.

4. **Supply chain transparency:**
   - Blockchain can increase transparency in multi-party, complicated supply chains by tracking each step along the way. As a result, confidence is increased and product provenance and authenticity are guaranteed.
   - IoT devices with sensors and RFID tags can instantly update the blockchain with information on the movement and state of the items.

5. **Traceability and quality assurance:**
   - Each IoT device's history, including its production, testing, and maintenance, may be accurately and permanently recorded using the immutable ledger offered by blockchain technology.
   - This traceability may improve quality control and aid in the speedy discovery of malfunctioning network elements or devices.

6. **Enhanced data integrity:**
   - Data from IoT devices may be cryptographically signed before being stored on a blockchain. This guarantees that the origin and integrity of the data are verified.
   - The blockchain's decentralized structure precludes a single point of failure that can jeopardize data integrity [6].

7. **Distributed analytics and insights:**
   - While ensuring data privacy, IoT-generated data may be analyzed on the blockchain itself or via off-chain processing.
   - Distributed analytics may provide in-the-moment understanding and forecasting without jeopardizing the security of the unprocessed data.

8. **Automating transactions and processes:**
   - In IoT contexts, smart contracts may automate a variety of operations, such as the automated reordering of goods when inventory levels are low.
   - By eliminating the need for human involvement, these automated procedures help to eliminate mistakes and delays.

9. **Immutable audit trails:**
   - The openness and immutability of blockchain technology may provide a trustworthy audit trail for reasons of regulatory compliance and accountability.
   - All alterations or interactions inside the network may be tracked back to their source, guaranteeing accountability and transparency.

IoT ecosystems' interoperability and trust might be greatly improved by combining blockchain technology with ubiquitous computing. Blockchain technology can assist in resolving issues with identity management, data sharing, security, and other issues by offering a decentralized, secure, and standardized basis, eventually encouraging a more connected and dependable IoT ecosystem.

The Internet of Things (IoT) and ubiquitous computing bring possibilities and difficulties for many industries. Collaboration between industry stakeholders, governments, and researchers is necessary to address the issues and take advantage of the opportunities while minimizing the hazards associated with this technology.

12.3 INTEGRATION OF BLOCKCHAIN IN UBIQUITOUS COMPUTING

The Internet of Things (IoT), also known as ubiquitous computing, and blockchain technology may be used to create groundbreaking solutions that improve security, privacy, trust, and interoperability across IoT ecosystems. Blockchain may be incorporated into ubiquitous computing in the following ways.

12.3.1 Use cases and applications

Numerous use cases and applications that make the most of the advantages of both technologies are made possible by the combination of blockchain technology with ubiquitous computing (IoT). Here are a few famous instances.

Supply chain management: Blockchain can improve supply networks' traceability and transparency. IoT devices with sensors and RFID tags can monitor the movement and condition of items. The blockchain may be used to track every stage of the supply chain, assuring authenticity and lowering the danger of counterfeiting.

1. **Smart cities:** Blockchain can power various aspects of smart cities, such as traffic management, waste management, energy distribution, and public services. IoT sensors in urban infrastructure can gather real-time data, which is then stored securely and transparently [7] on the blockchain.
2. **Healthcare and medical devices:** IoT devices in healthcare can monitor patients' vital signs, medication adherence, and health conditions. Blockchain ensures the security and privacy of sensitive patient data, while also enabling secure sharing of medical records among authorized parties.
3. **Energy management:** IoT devices can collect data about energy consumption and production. Blockchain can enable peer-to-peer

energy trading, where excess energy generated by one entity can be sold to others within a local energy community.

4. **Automotive industry:** Connected vehicles generate massive amounts of data related to performance, maintenance, and driving patterns. Blockchain can secure and authenticate this data, enabling tamper-proof records for vehicle history, insurance claims, and recalls.

5. **Agriculture and food safety:** IoT devices can monitor agricultural processes, ensuring proper conditions for crops and livestock. Blockchain can track the entire journey of food products from farm to table, reducing the risk of foodborne illnesses and ensuring food safety.

6. **Asset tracking and management:** IoT devices can track the location and condition of assets such as machinery, equipment, and high-value items. Blockchain's immutable ledger ensures accurate and auditable records of asset movement.

7. **Environmental monitoring:** IoT sensors can collect environmental data like air quality, water levels, and temperature. This data can be securely stored on the blockchain, providing reliable records for research, regulatory compliance, and environmental analysis.

8. **Retail and consumer goods:** IoT devices can enhance the retail experience by enabling personalized marketing, inventory management, and automatic replenishment of products. Blockchain ensures the authenticity of high-value items and reduces fraud.

9. **Wearable devices and health tracking:** Wearable IoT devices, such as fitness trackers and smartwatches, can record users' health data. This data can be securely stored on the blockchain, allowing users to have control over their health information and share it securely with healthcare providers.

10. **Home automation and security:** IoT devices in smart homes can control lighting, temperature, security systems, and more. Blockchain can enhance security by enabling decentralized authentication and secure communication among devices.

11. **Industrial IoT (IIoT):** In industrial settings, IoT devices monitor machines, processes, and equipment. Blockchain can enable secure data sharing between different stakeholders in the supply chain, enhancing collaboration and efficiency.

These use cases highlight the diverse ways in which blockchain can enhance the capabilities of ubiquitous computing. By providing data integrity, security, and transparency, blockchain can enable new levels of trust and interoperability in IoT ecosystems, paving the way for innovative applications across various industries.

12.3.2 Industry examples of blockchain and ubiquitous computing synergy

The synergy between blockchain technology and ubiquitous computing (IoT) offers significant benefits across various industries [8]. Here are some industry examples that showcase how the integration of blockchain and IoT can create innovative solutions:

1. **Supply chain and logistics:** Blockchain can enhance transparency and traceability in supply chains, while IoT devices track the movement and condition of goods. This synergy ensures that every step in the supply chain is recorded on an immutable ledger, reducing fraud, ensuring authenticity, and optimizing logistics processes.
2. **Smart cities:** Blockchain can power several features of smart cities, including traffic, trash, and energy management, as well as public services. Urban infrastructure IoT sensors may collect real-time data, which is subsequently safely and openly recorded on the blockchain.
3. **Medical devices and healthcare:** IoT devices in healthcare may track patients' vital signs, medication compliance, and health problems. Blockchain guarantees the confidentiality and privacy of private patient information while also facilitating safe record transfer between authorized parties.
   a. **Energy management:** IoT gadgets may gather information on energy generation and use. Peer-to-peer energy trading can be made possible via blockchain, allowing surplus energy produced by one entity to be sold to others in a nearby energy community.
   b. **The automotive sector:** Connected automobiles provide enormous volumes of data on operation, upkeep, and driving habits. Blockchain can verify and safeguard this data, providing tamper-proof records for insurance claims, recalls, and car history.
   c. **Agriculture and food safety:** IoT devices can keep an eye on agricultural procedures to ensure that crops and livestock are kept in the right conditions. By tracking food items' complete path from farm to table, blockchain can lower the risk of foodborne diseases and guarantee food safety.
   d. **Asset tracking and management:** IoT gadgets can monitor the whereabouts and state of assets, including machinery, equipment, and expensive goods. The unchangeable ledger of a blockchain provides precise and auditable records of asset movement.
   e. **Environmental monitoring:** IoT sensors may gather information about the environment, including the temperature, water levels, and air quality. The blockchain may be used to safely store this data, creating trustworthy records for research, legal compliance, and environmental studies.

f. **Retail and consumer goods:** By enabling personalized marketing, inventory management, and product restocking automatically, IoT devices may improve the retail experience. Blockchain prevents fraud and assures the validity of valuable commodities.

g. **Wearable device and health tracking:** Wearable IoT devices, such as fitness trackers and smartwatches, can capture information on a user's health. Users may own their health information and securely share it with healthcare professionals since this data can be safely maintained on the blockchain.

h. **Home automation and security:** IoT devices in smart homes may manage lighting, heating, security systems, and more. By providing decentralized authentication and safe device communication, blockchain can improve security.

i. **Industrial IoT (IIoT):** IoT devices are used to monitor machinery, processes, and equipment in industrial environments. Blockchain can improve collaboration and efficiency in the supply chain by enabling safe data sharing between many parties.

The various ways that blockchain might improve the capabilities of ubiquitous computing are highlighted by these use cases. Blockchain can allow unprecedented levels of trust and interoperability in IoT ecosystems by supplying data integrity, security, and transparency, opening the door for creative applications in several sectors.

These examples show how the integration of blockchain and IoT may revolutionize a number of sectors by enhancing trust, efficiency, transparency, and security. These technologies have a significant potential to alter established business procedures and open up new possibilities as they develop further.

12.3.3 Smart cities and IoT device management

Smart cities use Internet of Things (IoT) technology to optimize resource allocation, enhance urban infrastructure, and improve citizen quality of life. Smart city IoT devices are essential for data collection, system monitoring, and data-driven decision-making. To guarantee the efficient running of smart city programs, these IoT devices must be managed effectively. The following describes how smart cities manage IoT devices:

1. **Device provisioning:** The first step in creating a smart city is to configure and deploy IoT devices across the built environment. Setting up the basic setups and installing sensors are examples of jobs in this category.

2. **Connectivity management:** For IoT devices, dependable connectivity is crucial. Depending on the needs and locations of the devices, smart cities use a variety of connection solutions, including cellular, Wi-Fi, and low-power networks like LoRaWAN.

3. **Device maintenance and monitoring:** Smart cities regularly check on the functionality and health of IoT devices. This entails monitoring the connection, battery life, and other factors of the gadget. Automated warnings and alerts aid in the rapid identification of problems for maintenance.

4. **Firmware and software upgrades:** To enhance functionality, security, and performance, IoT devices need frequent upgrades. To make sure that devices are running with the most recent firmware and software, smart cities remotely control and apply these upgrades.

5. **Data gathering and analysis:** IoT devices produce a lot of data. To acquire insights into urban patterns, including traffic flow, energy consumption, trash management, and public service use, smart cities gather, store, and analyze this data.

6. **Security and privacy:** The security and privacy of citizen data are top priorities in smart cities. Devices include security features like authentication and encryption to protect data from unauthorized access. Additionally, privacy laws are followed, guaranteeing the protection of residents' private information.

7. **Energy efficiency:** Especially for battery-powered IoT devices, controlling energy usage is essential. By installing energy-saving measures and planning data flows to minimize energy drain, smart cities optimize their use of energy.

8. **Scalability and interoperability:** Managing a large number of various IoT devices gets challenging as smart city initiatives grow. Standardized protocols, such MQTT or CoAP, are used in smart cities to guarantee compatibility and smooth communication between devices made by various manufacturers.

9. **Data integration and visualization:** IoT platforms, often called dashboards, are centralized platforms that incorporate data from multiple IoT devices. City managers may use these tools to visualize data in real time, identify patterns, and make defensible choices.

10. **Predictive maintenance:** Smart cities may use predictive maintenance techniques by analyzing data from IoT devices. In order to minimize downtime, this entails recognizing probable device failures or performance deterioration before they happen.

IoT-enabled traffic sensors and cameras assist in managing traffic flow and maximizing parking. Traffic lights are changed in real time to control congestion and direct cars to open parking places.

Monitoring of the environment using IoT gadgets keeps an eye on pollution, noise levels, and air quality. Cities may use the acquired data to monitor environmental health, locate sources of pollution, and put improvement measures in place for the air.

Successful smart city implementations depend on efficient IoT device management. Smart cities may take use of the promise of IoT technology to build more sustainable, effective, and livable urban settings by ensuring that devices are well maintained, secure, and linked into a cohesive system.

12.3.4 Management of the supply chain and traceability

The combination of blockchain technology with IoT devices may have a substantial positive impact on supply chain management and traceability, improving transparency, security, and efficiency across the supply chain process. The following describes how various technologies interact in supply chain management:

1. **Transparency and traceability:** At different points throughout the supply chain, real-time data is gathered using IoT sensors and devices. As items travel from suppliers to manufacturers to distributors and retailers, these sensors can monitor their movement, temperature, humidity, and other environmental factors. An immutable and transparent record of each step in the supply chain is made using this data, which is safely stored on a blockchain.
2. **Origin and validity:** Blockchain makes it possible to verify a product's origin and validity. The blockchain keeps track of each item's path, including its origin, production procedures, and any previous phases. Authorized parties, including customers, may access this data to confirm the legitimacy of items.
3. **Counterfeit prevention:** Supply chains may more effectively battle fake products by merging IoT sensors with blockchain. By checking a product's blockchain history, counterfeit goods may be promptly located and taken off the market. Customers may also get comprehensive information about a product's validity by scanning QR codes or NFC tags on it.
4. **Recall management:** In the case of a product recall, blockchain technology and Internet of Things devices may be used to pinpoint the precise batch of afflicted goods. Instead of a general recall of all items, this focused strategy minimizes waste by selectively recalling the particular products that are at danger.
5. **Quality control:** IoT sensors can keep an eye on how goods are moved and kept. A violation in a set of parameters, such as temperature or humidity, results in the prompt recording of the information on the blockchain. By doing so, it is ensured that quality control procedures are followed and that any deviations are noted.
6. **Smart contracts for automation:** Supply chain procedures may be automated using smart contracts. For instance, payments may be automatically triggered when precise criteria are satisfied (such as cargo arriving at a particular area), eliminating the need for middlemen and delays.

7. **Supplier verification:** By safely storing details about a supplier's certificates, regulatory compliance, and prior business dealings, blockchain helps authenticate and verify suppliers. By doing this, the process of supplier verification is streamlined, and the danger of dealing with untrustworthy partners is reduced.
8. **Environmental impact tracking:** IoT devices can keep track of how supply chain operations affect the environment, including things like energy use and carbon emissions. To quantify and show the supply chain's sustainability, this data may be kept on a blockchain.
9. **Audit and compliance:** Due to the transparency and immutability of the blockchain, audits and compliance checks are more accurate and efficient. The blockchain may be accessed by regulatory bodies and auditors to confirm adherence to set norms and laws.
10. **Real-time visibility:** IoT devices provide users access to real-time information on the location and condition of their items. Since this data is maintained on the blockchain, all interested parties have access to accurate and current information on the state of the supply chain.

The symbiotic relationship between blockchain technology and IoT devices transforms supply chain management by increasing efficiency, transparency, and traceability. These technologies make it possible to trace goods precisely, lower the chance of fraud and counterfeiting, and increase the transparency and accountability of the whole supply chain.

12.4 CONCLUSION

The use of blockchain technology in ubiquitous computing offers the potential to improve data integrity, interoperability, and security, among other benefits. Nevertheless, its implementation has to take into account the particular use cases and strike a balance between the advantages and the related challenges. The choice to use blockchain technology needs to be predicated on a meticulous analysis of the one-of-a-kind necessities of each application, with consideration given to aspects such as scalability, energy efficiency, and intricacy. Blockchain may not be a one-size-fits-all solution for all elements of ubiquitous computing, despite the fact that it has the potential to give helpful answers in a variety of use cases.

REFERENCES

1. Mediatized Worlds: Culture and Society in a Media Age, Cultural Studies, Media and Communication, Regional Cultural Studies, ISBN-978-1-137-30034-8 Published: 12 March 2014, doi: 10.1057/9781137300355
2. https://www.geeksforgeeks.org/blockchain-technology-introduction/?cv=1

3. Ramander Singh, Rajesh Kumar Tyagi, Anil Kumar Mishra and Umakant Choudhury (2023). Review on Applicability and Utilization of Blockchain Technology in Ubiquitous Computing, Recent Advances in Computer Science and Communications, 16(7), e210323214821. doi: 10.2174/266625581 6666230321120653
4. A. Abouaomar, S. Cherkaoui, Z. Mlika and A. Kobbane (2021). Resource provisioning in edge computing for latency-sensitive applications, IEEE Internet of Things Journal, 8(14), 11088–11099, doi: 10.1109/JIOT.2021. 3052082, doi: 10.1007/978-3-319-99061-3_1
5. Hany F. Atlam and Gary B. Wills (2019). Chapter One – Technical Aspects of Blockchain and IoT. In: Shiho Kim, Ganesh Chandra Deka, Peng Zhang (eds.), Advances in Computers, vol. 115, , pp. 1–39, Elsevier, ISSN 0065-2458, ISBN 9780128171899, doi: 10.1016/bs.adcom.2018.10.006
6. Sachin Shetty, Charles A. Kamhoua and Laurent L. Njilla (eds.) (2019). Blockchain for Distributed Systems Security. John Wiley & Sons.
7. K. Biswas and V. Muthukkumarasamy (2016). Securing Smart Cities Using Blockchain Technology, 2016 IEEE 18th International Conference on High Performance Computing and Communications; IEEE 14th International Conference on Smart City; IEEE 2nd International Conference on Data Science and Systems (HPCC/SmartCity/DSS), Sydney, NSW, Australia, pp. 1392–1393, doi: 10.1109/HPCC-SmartCity-DSS.2016.0198
8. N.K. Savelyeva, A.A. Semenova, L.V. Popova and L.V. Shabaltina (2022). Smart Technologies in Agriculture as the Basis of Its Innovative Development: AI, Ubiquitous Computing, IoT, Robotization, and Blockchain. In: E.G. Popkova and B.S. Sergi (eds.), Smart Innovation in Agriculture. Smart Innovation, Systems and Technologies, vol. 264. Springer. doi: 10.1007/ 978-981-16-7633-8_4

Index

For Product Safety Concerns and Information please contact our EU
representative GPSR@taylorandfrancis.com
Taylor & Francis Verlag GmbH, Kaufingerstraße 24, 80331 München, Germany

www.ingramcontent.com/pod-product-compliance
Lightning Source LLC
Chambersburg PA
CBHW060555220326
41598CB00024B/3109

9 781032 609881